THIRD EDITION

Statistics:
A Guide to the Unknown

THE WADSWORTH & BROOKS/COLE STATISTICS/PROBABILITY SERIES

R. Becker, J. Chambers, A. Wilks, *The New S Language: A Programming Environment for Data Analysis and Graphics*
P. Bickel, K. Doksum, J. Hodges, Jr., *A Festschrift for Erich L. Lehmann*
G. Box, *The Collected Works of George E. P. Box, Volumes I and II,* G. Tiao, editor-in-chief
L. Breiman, J. Friedman, R. Olshen, C. Stone, *Classification and Regression Trees*
J. Chambers, W. S. Cleveland, B. Kleiner, P. Tukey, *Graphical Methods for Data Analysis*
W. S. Cleveland, M. McGill, *Dynamic Graphics for Statistics*
K. Dehnad, *Quality Control, Robust Design, and the Taguchi Method*
R. Durrett, *Lecture Notes on Particle Systems and Percolation*
F. Graybill, *Matrices with Applications in Statistics, Second Edition*
L. Le Cam, R. Olshen, *Proceedings of the Berkeley Conference in Honor of Jerzy Neyman and Jack Kiefer, Volumes I and II*
P. Lewis, E. Orav, *Simulation Methodology for Statisticians, Operations Analysts and Engineers*
H. J. Newton, *TIMESLAB*
J. Rawlings, *Applied Regression Analysis: A Research Tool*
J. Rice, *Mathematical Statistics and Data Analysis*
J. Romano, A. Siegel, *Counterexamples in Probability and Statistics*
J. Tanur, F. Mosteller, W. Kruskal, E. Lehmann, R. Link, R. Pieters, G. Rising, *Statistics: A Guide to the Unknown, Third Edition*
J. Tukey, *The Collected Works of J. W. Tukey,* W. S. Cleveland, editor-in-chief
Volume I: Time Series: 1949–1964, edited by D. Brillinger
Volume II: Time Series: 1965–1984, edited by D. Brillinger
Volume III: Philosophy and Principles of Data Analysis: 1949–1964, edited by L. Jones
Volume IV: Philosophy and Principles of Data Analysis: 1965–1986, edited by L. Jones
Volume V: Graphics: 1965–1985, edited by W. S. Cleveland

THIRD EDITION

Statistics:
A Guide to the Unknown

Edited by

Judith M. Tanur *State University of New York, Stony Brook*

Frederick Mosteller *Chairman Harvard University*

William H. Kruskal *University of Chicago*

Erich L. Lehmann *University of California, Berkeley*

Richard F. Link *Richard F. Link & Associates, Inc.*

Richard S. Pieters *Instructor Emeritus, Phillips Academy, Andover, MA*

Gerald R. Rising *State University of New York, Buffalo*

The Joint Committee on the Curriculum in Statistics and Probability of the American Statistical Association and the National Council of Teachers of Mathematics

Wadsworth & Brooks/Cole Advanced Books & Software
Pacific Grove, California

Wadsworth & Brooks/Cole Advanced Books & Software
A Division of Wadsworth, Inc.

Printed in the United States of America

10 9 8 7 6 5 4 3

Library of Congress Cataloging in Publication Data

Statistics: a guide to the unknown/edited by Judith M. Tanur; and by Frederick Mosteller . . . [et al.] (The Joint Committee on the Curriculum in Statistics and Probability of the American Statistical Association and the National Council of Teachers of Mathematics).—3rd ed.
p. cm.
Includes bibliographies and index.
ISBN 0-534-09492-9
1. Mathematical statistics. I. Tanur, Judith M. II. Mosteller, Frederick. III. Joint Committee on the Curriculum in Statistics and Probability.
QA276.16.S84 1988
519.5—dc19 88-25889
CIP

Sponsoring Editor: *John Kimmel*
Editorial Assistant: *Maria Tarantino*
Production Editor: *Linda Loba*
Manuscript Editor: *Betty Berenson*
Permissions Editor: *Carline Haga*
Cover and Interior Design: *Flora Pomeroy*
Cover Photo: *Jerome Friar*
Art Coordinator: *Lisa Torri*
Interior Illustration: *Judith Macdonald; Art by Ayxa*
Typesetting: *Bookends Typesetting*
Cover Printing: *Phoenix Color Corporation*
Printing and Binding: *Arcata Graphics-Fairfield*

Preface

To prepare a volume describing important applications of statistics and probability in many fields of endeavor—this was the project that the American Statistical Association and the National Council of Teachers of Mathematics (ASA–NCTM) Committee invited me to help with in early 1969. It was the Committee's view that more statistics and its background in probability would be desirable in the school curriculum; thus it would be desirable to show how broadly these tools are applied. The Committee planned this book primarily for readers without special knowlege of statistics, probability, or mathematics. This audience includes especially parents of schoolchildren, school superintendents, principals, and board members, but also teachers of mathematics and their supervisors, and, finally, young people themselves. *Statistics: A Guide to the Unknown* is the result. During the time of the book's preparation several of us who were working on it and teaching simultaneously found much of the material very useful—even inspirational—to undergraduate and graduate students. Unexpectedly, the book had an additional function as an auxiliary textbook, and it has frequently been used in that way.

Instead of teaching technical methods, the essays illustrate past accomplishments and current uses of statistics and probability. In choosing the actual essays to include, the Committee and I aimed at illustrating a wide variety of fields of application, but we did not attempt the impossible task of covering all possible uses. Even in the fields included, attempts at complete coverage have been deliberately avoided. We discouraged authors from writing essays that could be entitled "All Uses of Statistics in . . . ." Rather, we asked authors to stress one or a very few important problems within their field of application and to explain how statistics and probability help to solve them and why the solutions are useful to the nation, to science, or to the people who originally posed the problem. In the past, for those who were unable to cope with very technical material, such essays had been hard to find.

To us, this spread of applications has renewed our appreciation of the unity in diversity that is statistics. On the one hand, we found the same, or similar, statistical techniques being applied to unrelated fields. Authors describe the use of correlation and regression in contexts as diverse as a study of the sun, the fair grading of tests, the effects of taxation on cigarette smoking, and an investigation of the effects of a crackdown on speeders. Other authors deal with applications of sampling theory in such disparate fields as accounting, improving the U.S. Census, and estimating the size of whale populations. And essay after essay discusses experimental design and the necessity, as well as the difficulty, of making inferences from less-than-perfect data. Certainly this unity in diversity helps to demonstrate to the general public the wide usefulness of statistical tools.

On the other hand, we found the essays could be grouped into unities of subject matter with differing statistical techniques. For example, two otherwise unlike essays deal with creating customer satisfaction with a manufactured product. At least five essays describe very different methods of studying diseases, their causes, and cures—the testing of the value of the Salk vaccine, an explanation of the uses of twins in research on illness, the use of randomization to study the effects of innovations in hospitals, a study of the possible deleterious effects of an anesthetic, and a study of health insurance and the effects of decreasing expenditures on health outcomes.

Once the essays had been assembled and edited, we had to decide on their order. Several orderings seemed feasible: We might group the essays by type of statistical tools employed, thus stressing the unity of statistical tools and ignoring the diversity of usual disciplinary lines; we might group the essays by the method used for collecting data—sample survey, experiment, Census material, and so on; or we might group them by the subject matter of the application.

What we have chosen is the last of these modes of organization. We have classified the essays into four broad areas by field of application, with subdivisions within some. Each subdivision is small enough, cohesive enough, and digestible enough to be read as a single unit and to give an overview of applications within a narrow field. But we were unwilling to forgo the advantages of the other possible methods of classification; following the main table of contents, therefore, are two alternate tables of contents, the first organized by the method of collecting the data and the second by statistical tools. In the latter listing, an essay has been listed under a heading whenever the author used that tool, or whenever we felt the reader might learn something about the technique by looking at the essay, or both.

These efforts at classifying emphasized, for us, an aspect of the book we had not deliberately planned or even been aware of earlier. It turned out that we had a large group of essays dealing with public policy, many of them classified under our main grouping entitled "Our Social World." We also found that several essays in this group deal with the evaluation of reforms or changes in policy. On the one hand, we found ourselves with descriptions of two large-scale field experiments: the health insurance experiment and the Salk vaccine trials. It seems that in the United States until recently, we have done few of these controlled experiments, and it appears to the Committee that one of the jobs that statisticians have been somewhat neglecting is explaining to the public the possibilities and values of experimentation. The public needs such explanations to have a sound basis for deciding whether it wants such experimentation to be carried out. On the other hand, several articles deal with nonexperimental (or quasi-experimental) evaluations of reforms: Did the Connecticut crackdown on speeding decrease traffic accidents? Is a particular anesthetic dangerous? Is a hiring policy discriminatory? Does increased taxation deter teenagers from smoking?

We hope that both types of essays will contribute to a greater appreciation of how hard it is to find out whether a program is accomplishing its purposes. Such understanding would give people a little more sympathy for government officials who are trying to do difficult jobs under severe handicaps. It may also, as pointed out above, encourage them to press government to do better-controlled field studies both in advance of and while instituting social reforms.

There is an old saw that a camel is a horse put together by a committee. Our authors supplied exceedingly well formed and attractive anatomical parts, but to the extent that this book gaits well, credit is due primarily to a most talented and dedicated Committee. In general, the approach to unanimity in the Committee members' critical review of and suggestions about the essays was phenomenal. And, though they may occasionally have been divided about the strong and weak points of a particular essay, they were constantly united in their purpose of producing a useful book and in their ability to find something more than 24 hours a day to work on it. This dedication, together with my own compulsiveness, has undoubtedly created difficulties for our authors. Nevertheless, our authors persevered and deserve enormous thanks from me, from the Committee, and from the statistical profession at large.

A work of art or literature may be ageless—a work of science is not. It has been 16 years since the first edition of *Statistics: A Guide to the Unknown* (SAGTU) appeared, and 10 years since the second edition. Much has happened in the world of statistics in that time, and there have been major efforts to introduce statistics to the general public via books and even television programs. Indeed, I have recently written a preface for a Chinese SAGTU edition. So we felt it was time for a new and leaner edition of SAGTU.

The second edition expanded the first by adding two new essays and by appending study materials to the essays. The third edition represents a major change. Authors have updated the essays that have been carried over from the earlier editions, and the book contains a dozen new essays. All of the essays continue to have study material.

Our thanks go to the Sloan Foundation whose grant made it possible to put the first edition together.

For the first edition there are others to thank as well: for the hard work and advice of George E. P. Box, Leo Breiman, Churchill Eisenhart, Thomas Henson, J. W. Tukey, and the late W. J. Youden; to the office of the American Statistical Association (and, in particular, to Edgar Bisgyer and John Lehman) for invaluable help in all the administrative work necessary to get out a book such as this; and similar thanks to the administration of the National Council of Teachers of Mathematics; to Edward Millman for careful and imaginative editorial assistance; and to other people at Holden-Day, especially Frederick H. Murphy, Walter Sears, and Erich Lehmann, our series editor; to Mrs. Holly Grano for acting as a long-distance and long-haul secretary; and to the many friends and colleagues both of the editor and of the Committee members who so often acted as unsung, but indispensable advisors.

As time has passed we continue to feel gratitude to all these people, especially to Frederick Murphy at Holden-Day who continued to watch over SAGTU for many years. The supplementary study material for the essays in the second edition was the invaluable contribution of David Lane, Donna and Leland Neuberg, Rick Persons, Haiganoush Preisler, and Esther Sid. It would have been impossible to produce this third edition without the invaluable assistance of Marjorie Olson. Betty Pond also deserves my thanks for taking care of many details, and we are most grateful for the assistance of John Kimmel of Wadsworth and Linda Loba of Brooks/Cole in the production process. But, once again, our greatest thanks go to the authors of the essays, who produced fine manuscripts under severe time pressures and were patient and responsive in the face of our constant requests for revisions.

It is my hope, and the hope of all of the editorial board, that SAGTU will continue to be useful as a supplementary text and will continue to do missionary work encouraging the inclusion of statistics and probability in curricula.

Judith M. Tanur
Great Neck, NY
February 4, 1988

Foreword to the Third Edition

The origins of this work are rooted in the great change and advance in mathematics education initiated in 1954 when the Commission on Mathematics of the College Entrance Examination Board brought together, for a sustained study of the curriculum, teachers and administrators of mathematics from several sources: secondary schools, teachers' colleges, and colleges and universities. Prior to that gathering, the several groups of teachers had seldom worked together on the curriculum. That meeting of minds has developed and continued in many directions; one of its long-run consequences was the establishment of the Joint Committee of the American Statistical Association (ASA) and the National Council of Teachers of Mathematics (NCTM) on the Curriculum in Statistics and Probability. By late 1967, such cooperation between school and college teachers was widespread, and it was easy for Donovan Johnson, then president of NCTM, and me, then president of ASA, to set up the Joint Committee to review matters in the teaching of statistics and probability.

The purpose of the Joint Committee is to encourage the teaching of statistics in schools. The sponsoring societies appropriately address such teaching because statistics is a part of the mathematical sciences that deals with many practical, as well as esoteric, subjects and is especially organized to treat the uncertainties and complexities of life and society. To explain why more statistics need to be taught, the Joint Committee felt that it had to make clear to the public what sorts of contributions statisticians make to society.

When describing work in the mathematical sciences, one must make a major decision as to what level of mathematics to ask of the reader. Although the Joint Committee serves professional organizations whose subject matter is strongly mathematical, we decided to explain statistical ideas and contributions without dwelling on their mathematical aspects. This was a bold stroke, and our authors were surprised that we largely held firm.

The Joint Committee has been extremely fortunate to find so many distinguished scholars willing to participate in this educational project. The authors' reward is almost entirely in their contribution to the appreciation of statistics. We have been fortunate, too, to have Judith Tanur as editor of the collection and hard-working committee members as her staff.

In a parallel writing effort, the Joint Committee has also produced a series of pamphlets for classroom teaching entitled *Statistics by Example.* Intended for students whose mathematical preparation is modest, these volumes teach statistics by means of real life examples. That effort differs from this one in that the student learns specific techniques, tools, and concepts by starting from concrete examples. (The publisher is Addison-Wesley, Sand Hill Road, Menlo Park, CA 94025).

The Joint Committee continues active to this day. Among its recent activities is a National Science Foundation–sponsored program entitled "Quantitative Literacy," which aims to educate teachers to train other teachers to teach statistics.

Some readers may wish to know how to become statisticians, and others may have the obligation to advise students about career opportunities. The brochure *Careers in Statistics* (obtainable from the American Statistical Association, 1429 Duke Street, Alexandria, VA 22314-3402) provides information about the nature of the work and the training required for various statistical specialties.

What made this effort possible was an initial grant from the Sloan Foundation. Over the years we have also benefited from a number of courtesies extended by the Russell Sage Foundation and by the Social Science Research Council. The national offices of the ASA and the NCTM have been helpful through three editions. The original publisher, Fred Murphy of Holden-Day, Inc., gave us attractive publications through two editions, and John Kimmel of Wadsworth and Brooks/Cole has continued this tradition. Marjorie Olson managed the manuscript for this edition, and Cleo Youtz contributed to its preparation.

Finally, we have no monopoly on the task of explaining statistics to the public. We urge others to provide their views on the purposes, the methods, and the results of statistical science. We happily note that some have done so.

Frederick Mosteller, Chairman
Editorial Committee of
Statistics: A Guide to the Unknown
Cambridge, Massachusetts
February 14, 1988

Contents

PART THREE
Our Social World

Communicating with Others

People at Work

People at School and Play

Counting People and Their Goods

PART FOUR
Our Physical World

Essays Classified by Data Sources

SAMPLES

Of Individuals and Their Aggregations

Of Records and the Physical Universe

AVAILABLE DATA

SURVEYS AND QUESTIONNAIRES

EXPERIMENTS

QUASI-EXPERIMENTS

OBSERVATIONAL STUDIES

Essays Classified by Statistical Tools

ESTIMATION, HYPOTHESIS TESTING, BAYESIAN AND EMPIRICAL BAYESIAN ANALYSIS, AND DATA ANALYSIS

Estimation

Hypothesis Testing

Bayesian and Empirical Bayesian Analysis

Data Analysis

TABLES, GRAPHS, AND MAPS

Tables

Graphs and Maps

PERCENTS AND RATES, STANDARDIZATION AND ADJUSTMENT

Percents and Rates

Standardization and Adjustment

TIME SERIES AND INDEX CONSTRUCTION

PROBABILITY AND MODELING

Probability

Modeling

SAMPLING AND RANDOMIZATION

Sampling

Randomization

CORRELATION AND REGRESSION

MULTIDIMENSIONAL SCALING AND DISCRIMINANT ANALYSIS

Multidimensional Scaling

Discriminant Analysis

TESTS AND MEASUREMENTS

FORECASTING AND PREDICTION

DECISION MAKING

THIRD EDITION

Statistics:
A Guide to the Unknown

PART ONE
Our Biologic World

Health and Sickness

People and Animals

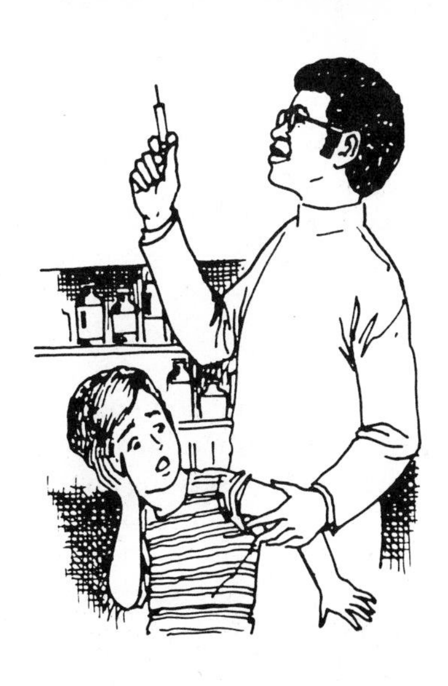

The Biggest Public Health Experiment Ever: The 1954 Field Trial of the Salk Poliomyelitis Vaccine

Paul Meier *University of Chicago*

The largest and, until the 1980's, the most expensive medical experiment in history was carried out in 1954. Well over a million young children participated, and the immediate direct costs were over 5 million midcentury dollars. The experiment was carried out to assess the effectiveness, if any, of the Salk vaccine as a protection against paralysis or death from poliomyelitis. The study was elaborate in many respects, most prominently in the use of placebo controls (children who were inoculated with simple salt solution) assigned at random (that is, by a carefully applied chance process that gave each volunteer an equal probability of getting vaccine or salt solution) and subjected to a double-blind evaluation (that is, an arrangement under which neither the children nor the physicians who evaluated their subsequent state of health knew who had been given the vaccine and who the salt solution).

Why was such elaboration necessary? Did it really result in more or better knowledge than could have been obtained from much simpler studies? These are the questions on which this discussion is focused.

BACKGROUND

Polio was never a common disease, but it certainly was one of the most frightening and, in many ways, one of the most inexplicable in its behavior. It struck hardest at young children, and, although it was responsible for only about 6% of the deaths in the age group 5 to 9 in the early 1950s, it left many helpless cripples, including some who could survive only in a respirator. It appeared in epidemic waves, leading to summer seasons in which some communities felt compelled to close swimming pools and restrict public gatherings as cases increased markedly from week to week; other communities, escaping an epidemic one year, waited in trepidation for the year in which their turn would come. Rightly or not, this combination of selective attack upon the most helpless age group, and the inexplicable vagaries of its epidemic behavior, led to far greater concern about polio as a cause of death than other causes, such as auto accidents, which are more frequent and, in some ways, more amenable to community control.

The determination to mount a major research effort to eradicate polio arose in no small part from the involvement of President Franklin D. Roosevelt, who was struck down by polio when a successful young politician. His determination to overcome his paralytic handicap and the commitment to the fight against polio made by Basil O'Connor, his former law partner, enabled a great deal of attention, effort, and money to be expended on the care and rehabilitation of polio victims and—in the end, more importantly—on research into the causes and prevention of the disease.

During the course of this research, it was discovered that polio is caused by a virus and that three main virus types are involved. Although clinical manifestations of polio are rare, it was discovered that the virus itself was not rare, but common, and that most adults had experienced a polio infection sometime in their lives without ever being aware of it.

This finding helped to explain the otherwise peculiar circumstance that polio epidemics seemed to hit hardest those who were better off hygienically (that is, those who had the best nutrition, most favorable housing conditions, and were otherwise apparently most favorably situated). Indeed, the disease seemed to be virtually unknown in those countries with the poorest hygiene. The explanation is that because there was plenty of polio virus in the less-favored populations, almost every infant was exposed to the disease early in life while still protected by the immunity passed on from the mother. As a result, everyone had had polio, but under protected circumstances, and, thereby, everyone had developed immunity.

As with many other virus diseases, an individual who has been infected by polio and recovered is usually immune to another attack (at least by a virus strain of the same type). The reason for this is that the body, in fighting the infection, develops *antibodies* (a part of the gamma globulin fraction of the blood) to the *antigen* (the protein part of the polio virus). These antibodies remain in the bloodstream for years, and even when their level declines so far as to be scarcely measurable, there are usually enough of them to prevent a serious attack from the same virus.

Smallpox and influenza illustrate two different approaches to the preparation of an effective vaccine. For smallpox, which has long been controlled by a vaccine, we use for the vaccine a closely related virus, cowpox, which is ordinarily incapable of causing serious disease in humans, but which gives rise to antibodies that also protect against smallpox. (In a very few individuals this vaccine is capable of causing a severe, and occasionally fatal, reaction. The risk is small enough, however, so that before smallpox was conquered we did not hesitate to expose all our schoolchildren to it in order to protect them from smallpox.) In the case of influenza, however, instead of a closely related live virus, the vaccine is a solution of the influenza virus itself, prepared with a virus that has been killed by treatment with formaldehyde. Provided that the treatment is not too prolonged, the dead virus still has enough antigenic activity to produce the required antibodies so that, although it can no longer infect, it is sufficiently like the live virus to be a satisfactory vaccine.

For polio, both of these methods were explored. A live-virus vaccine would have the advantage of reproducing in the vaccinated individual and, hopefully, giving rise to a strong reaction that would produce a high level of long-lasting antibodies. With such a vaccine, however, there might be a risk that a vaccine virus so similar to the virulent polio virus could mutate into a virulent form and itself be the cause of paralytic or fatal disease. A killed-virus vaccine should be safe because it presumably could not infect, but it might fail to give rise to an adequate antibody response. These and other problems stood in the way of the rapid development of a successful vaccine. Some unfortunate prior experience also contributed to the cautious approach of the researchers. In the 1930s, attempts had been made to develop vaccines against polio; two of these were actually in use for a time. Evidence that at least one of these vaccines had been responsible for cases of paralytic polio soon caused both to be promptly withdrawn from use. This experience was very much in the minds of polio researchers, and they had no wish to risk a repetition.

Research to develop both live and killed vaccines was stimulated in the late 1940s by the development of a tissue culture technique for growing polio virus. Those working with live preparations developed harmless strains from virulent ones by growing them for many generations in suitable tissue culture media. There was, of course, considerable worry lest these strains, when used as a vaccine in humans, might revert to virulence and cause paralysis or death. (It's now clear that the strains developed are indeed safe—a live-virus preparation taken orally is the vaccine presently in widespread use throughout the world.)

Those working with killed preparations, notably Jonas Salk, had the problem of treating the virus (with formaldehyde) sufficiently to eliminate its infectiousness, but not so long as to destroy its antigenic effect. This was more difficult than expected, and some early lots of the vaccine proved to contain live virus capable of causing paralysis and death. There are statistical issues in the safety story (Meier, 1957), but our concern here is with the evaluation of effectiveness.

EVALUATION OF EFFECTIVENESS

In the early 1950s the Advisory Committee convened by the National Foundation for Infantile Paralysis (NFIP) decided that the killed-virus vaccine developed by Jonas Salk at the University of Pittsburgh had been shown to be both safe and capable of inducing high levels of the antibody in children on whom it had been tested. This made the vaccine a promising candidate for general use, but it remained to prove that the vaccine actually would prevent polio in exposed individuals. It would be unjustified to release such a vaccine for general use without convincing proof of its effectiveness, so it was determined that a large-scale "field trial" should be undertaken.

That the trial had to be carried out on a very large scale is clear. For suppose we wanted the trial to be convincing if indeed the vaccine were 50% effective (for various reasons, 100% effectiveness could not be expected). Assume that, during the trial, the rate of occurrence of polio would be about 50 per 100,000 (which was about the average incidence in the United States during the 1950s). With 40,000 in the control group and 40,000 in the vaccinated group, we would find about 20 control cases and about 10 vaccinated cases, and a difference of this magnitude could fairly easily be attributed to random variation. It would suggest that the vaccine might be effective, but it would not be persuasive. With 100,000 in each group, the expected numbers of polio cases would be 50 and 25, and such a result would be persuasive. In practice, a much larger study was clearly required because it was important to get definitive results as soon as possible, and if there were relatively few cases of polio in the test area, the expected number of cases might be well under 50. It seemed likely, also, for reasons we shall discuss later, that paralytic polio, rather than all polio, would be a better criterion of disease, and only about half the diagnosed cases are classified "paralytic." Thus the relatively low incidence of the disease, and its great variability from place to place and time to time, required that the trial involve a huge number of subjects—as it turned out, over a million.

THE VITAL STATISTICS APPROACH

Many modern therapies and vaccines, including some of the most effective ones such as smallpox vaccine, were introduced because preliminary studies suggested their value. Large-scale use subsequently provided clear evidence of efficacy. A natural and simple approach to the evaluation of the Salk vaccine would have been to distribute it as widely as possible, through the schools, to see whether the rate of reported polio was appreciably less than usual during the subsequent season. Alternatively, distribution might be limited to one or a few areas because limitations of supply would preclude effective coverage of the entire country. There is even a fairly good chance that were one to try out an effective vaccine against the common cold, convincing evidence might be obtained in this way.

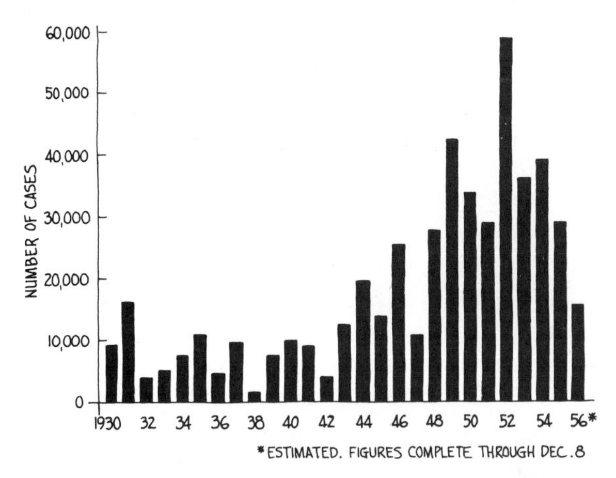

Figure 1 *Poliomyelitis in the U.S., 1930–56. Source: Rutstein (1957)*

In the case of polio—and, indeed, in most cases—so simple an approach would almost surely fail to produce clear-cut evidence. First, and foremost, we must consider how much polio incidence varies from season to season, even without any attempts to modify it. From Figure 1, which shows the annual reported incidence from 1930 through 1955, we see that had a trial been conducted in this way in 1931, the drop in incidence from 1931 to 1932 would have been strongly suggestive of a highly effective vaccine because the incidence dropped to less than a third of its previous level. Similar misinterpretations would have been made in 1935, 1937, and other years—for example, 1952. (On the general problem of drawing inferences from such time series data see the essay by Campbell.) One might suppose that such mistakes could be avoided by using the vaccine in one area, say, New York State, and comparing the rate of incidence there with that of an unvaccinated area, say, Illinois. Unfortunately, an epidemic of polio might well occur in Chicago—as it did in 1956—during a season in which New York had a very low incidence.

Another problem, more subtle, but equally burdensome, relates to the vagaries of diagnosis and reporting. There is no difficulty, of course, in diagnosing the classic respirator case of polio, but the overwhelming majority of cases are less clear-cut. Fever and weakness are common symptoms of many illnesses, including polio, and the distinction between weakness and slight transistory paralysis will be made differently by different observers. Thus the decision to diagnose a case as nonparalytic polio instead of some other disease might well be influenced by a physician's general knowledge or feeling about how widespread polio is in his or her community at the time.

These difficulties can be mitigated to some extent by setting down very precise criteria for diagnosis, but it is virtually impossible to obviate them completely when, as would be the case after the widespread introduction of a new vaccine, there is a marked shift in what the physician expects to find. This is most especially true when the initial diagnosis must be made by family physicians who cannot easily be indoctrinated in the use of a special set of criteria, as is the case with polio. Later evaluation by specialists cannot, of course, bring into the picture those cases originally diagnosed as something other than polio.

THE OBSERVED CONTROL APPROACH

The difficulties of the vital statistics approach were recognized by all concerned, and the initial study plan, although not judged entirely satisfactory, got around many of the problems by introducing a control group similar in characteristics to the vaccinated group. More specifically, the idea was to offer vaccination to all children in the second grade of participating schools and to follow the polio experience not only in these children but in the first- and third-grade children as well. Thus the vaccinated second-graders would constitute the *treated* group, and the first- and third-graders would constitute the *control* group. This plan follows what we call the *observed control approach.*

It is clear that this plan avoids many of the difficulties listed above. The three grades all would be drawn from the same geographic location so that an epidemic affecting the second grade in a given school would certainly affect the first and third grades as well. Of course, all subjects would be observed concurrently in time. The grades, naturally, would be different ages, and polio incidence does vary with age. Not much variation from grade to grade was expected, however, so it seemed reasonable to assume that the average of first and third grades would provide a good control for the second grade.

Despite the relative attractiveness of this plan and its acceptance by the NFIP advisory committee, serious objections were raised by certain health departments that were expected to participate. In their judgment, the results of such a study were likely to be insufficiently convincing for two important reasons. One is the uncertainty in the diagnostic process mentioned earlier and its liability to influence by the physician's expectations, and the other is the selective effect of using volunteers.

Under the proposed study design, physicians in the study areas would have been aware of the fact that only second-graders were offered vaccine, and in making a diagnosis for any such child, they would naturally and properly have inquired whether the child had been vaccinated. Any tendency to decide a difficult diagnosis in favor of nonpolio when the child was known to have been vaccinated would have resulted in a spurious piece of evidence favoring the vaccine. Whether or not such an effect was really operating would have been almost impossible to judge with assurance, and the results, if favorable, would have been forever clouded by uncertainty.

A less conjectural difficulty lies in the difference between those families who volunteer their children for participation in such a trial and those who do not. Not at all surprisingly, it was later found that those who do volunteer tend to be better educated and, generally, more well-to-do than those who do not participate. There was also evidence that those who agree to participate tend to be absent from school with a noticeably higher frequency than others. The direction of effect of such selection on the incidence of diagnosed polio is by no means clear before the fact, and this important difference between the treated group and the control group also would have clouded the interpretation of the results.

RANDOMIZATION AND THE PLACEBO CONTROL APPROACH

The position of critics of the NFIP plan was that the issue of vaccine effectiveness was far too important to be studied in a manner that would leave uncertainties in the minds of reasonable observers. No doubt, if the vaccine should appear to have fairly high effectiveness, most public health officials and the general public would accept it, despite the reservations. If, however, the observed control scheme were used, a number of qualified public health scientists would have remained unconvinced, and the value of the vaccine would be uncertain. Therefore, the critics proposed that the study be run as a scientific experiment with the use of appropriate randomizing procedures to assign subjects to treatment or to control and with a maximum effort to eliminate observer bias. This plan follows what we call the *placebo control approach.*

The chief objection to this plan was that parents of schoolchildren could not reasonably be expected to permit their children to participate in an experiment in which they might be getting only an ineffective salt solution instead of a probably helpful vaccine. It was argued further that the injection of placebo might not be ethically sound since a placebo injection carries a small risk, especially if the child unknowingly is already infected with polio.

The proponents of the placebo control approach maintained that, if properly approached, parents *would* consent to their children's participation in such an experiment, and they judged that because the injections would not be given during the polio season, the risk associated with the placebo injection itself was vanishingly small. Certain health departments took a firm stand: they would participate in the trial only if it were such a well-designed experiment. The consequence was that in approximately half the areas, the randomized placebo control method was used, and in the remaining areas, the alternating-grade observed control method was used.

A major effort was put forth to eliminate any possibility of the placebo control results being contaminated by subtle observer biases. The only firm way to accomplish this was to ensure that neither the subject, nor the parents, nor the diagnostic personnel could know which children had gotten the vaccine until all diagnostic decisions had been made. The method for achieving this result was to prepare placebo material that looked just like the vaccine but was without any antigenic activity, so that the controls might be inoculated and otherwise treated in just the same fashion as were the vaccinated.

Each vial of injection fluid was identified only by a code number so that no one involved in the vaccination or the diagnostic evaluation process could know which children had gotten the vaccine. Because no one knew, no one could be influenced to diagnose differently for vaccinated cases and for controls. An experiment in which both the subject getting the treatment and the diagnosticians who will evaluate the outcome are kept in ignorance of the treatment given each individual is called a *double-blind* experiment. Experience in clinical research has shown the double-blind experiment to be the only satisfactory way to avoid potentially serious observer bias when the final evaluation is in part a matter of judgment.

For most of us, it is something of a shock to be told that competent and dedicated physicians must be kept in ignorance lest their judgments be colored by knowledge of treatment status. We should keep in mind that it is not deliberate distortion of findings by the physician that concern the medical experimenter. It is rather the extreme difficulty in many cases of making an uncertain decision that, experience has shown, leads the best of investigators to be subtly influenced by information of this kind. For example, in the study of drugs used to relieve postoperative pain, it has been found that it is quite impossible to get an unbiased judgment of the quality of pain relief, even from highly qualified investigators, unless the judge is kept in ignorance of which patients were given the drugs.

The second major feature of the experimental method was the assignment of subjects to treatments by a careful randomization procedure. As we observed earlier, the chance of coming down with a diagnosed case of polio varies with a great many factors including age, socioeconomic status, and the like. If we were to make a deliberate effort to match up the treatment and control groups as closely as possible, we should have to take care to balance these and many other factors, and, even so, we might miss some important ones. Therefore, perhaps surprisingly, we leave the balancing to a carefully applied equivalent of coin tossing: we arrange that each individual has an equal chance of getting vaccine or placebo, but we eliminate our own judgment entirely from the individual decision and leave the matter to chance.

The gain from doing this is twofold. First, a chance mechanism usually will do a good job of evening out all the variables—those we didn't recognize in advance as well as those we did recognize. Second, if we use a chance mechanism in assigning treatments, we may be confident about the use of the theory of chance (that is, probability theory) to judge the results. We can then calculate the probability that so large a difference as that observed could reasonably be due solely to the way in which subjects were assigned to treatments, or whether, on the contrary, it is really an effect due to a true difference in treatments.

To be sure, there are situations in which a skilled experimenter can balance the groups more effectively than a random-selection procedure typically would. When some factors may have a large effect on the outcome of an experiment, it may be desirable, or even necessary, to use a more complex experimental design that takes account of these factors. However, if we intend to use probability theory to guide us in our judgment about the results, we can be confident about the accuracy of our conclusions only if we have used randomization at some appropriate level in the experimental design.

The final determinations of diagnosed polio proceeded along the following lines. All cases of poliolike illness reported by local physicians were subjected to special examination, and a report of history, symptoms, and laboratory findings was made. A special diagnostic group then evaluated each case and classified it as nonpolio, doubtful polio, or definite polio. The last group was subdivided into nonparalytic and paralytic, with paralytic divided into nonfatal and fatal polio. Only after this process was complete was the code broken and identification made for each case as to whether vaccine or placebo had been administered.

RESULTS OF THE TRIAL

The main results are shown in Table 1, which shows the size of the study populations, the number of cases classified as polio, and the disease rates; that is, the number of cases per 100,000 population. For example, the second line shows that in the placebo control area there were 428 reported cases, of which 358 were confirmed as polio, and, among these, 270 were classified as paralytic (including 4 that were fatal). The third and fourth rows show corresponding entries for those who were vaccinated and those who received placebo, respectively. Beside each of these numbers is the corresponding rate. Using the simplest measure—all reported cases—the rate in the vaccinated group is seen to be half that in the control group (compare the boxed rates in Table 1) for the placebo control areas. This difference is greater than could reasonably be ascribed to chance, according to the appropriate probability calculation. The apparent effectiveness of the vaccine is more marked as we move from reported cases to paralytic cases to fatal cases, but the numbers are small and it would be unwise to make too much of the apparent very high effectiveness in protecting against fatal cases. The main point is that the vaccine was a success; it demonstrated sufficient effectiveness in preventing serious polio to warrant its introduction as a standard public health procedure.

Not surprisingly, the observed control area provided results that were, in general, consistent with those found in the placebo control area. The volunteer effect discussed earlier, however, is clearly evident (note that the rates for those not inoculated differ from the rates for controls in both areas). Were the observed control information alone available, considerable doubt would have remained about the proper interpretation of the results.

Although there had been wide differences of opinion about the necessity or desirability of the placebo control design before, there was great satisfaction with the method after the event. The difference between the two groups, although substantial and definite, was not so large as to preclude doubts had there been no placebo controls. Indeed, there were many surprises in the more detailed data. It was known, for example, that some lots of vaccine had greater antigenic power than did others, and it might be supposed that they should have shown a greater protective effect. This was not the case; lots judged inferior

Table 1 Summary of study cases by diagnostic class and vaccination status (rates per 100,000)

				Poliomyelitis Cases									
		All Reported Cases		*Total*		*Paralytic*		*Non-paralytic*		*Fatal polio*		*Not Polio*	
Study Group	*Study Population*	*No.*	*Rate*	*No.*	*Rate*	*No.*	*Rate*	*No.*	*Rate*	*No.*	*Rate*	*No.*	*Rate*
All areas: Total	1,829,916	1,013	55	863	47	685	37	178	10	15	1	150	8
Placebo control areas: Total	749,236	428	57	358	48	270	36	88	12	4	1	70	9
Vaccinated	200,745	82	41	57	28	33	16	24	12	—	—	25	12
Placebo	201,229	162	81	142	71	115	57	27	13	4	2	20	10
Not inoculated*	338,778	182	54	157	46	121	36	36	11	—	—	25	7
Incomplete vaccinations	8,484	2	24	2	24	1	12	1	12	—	—	—	—
Observed control areas: Total	1,080,680	585	54	505	47	415	38	90	8	11	1	80	7
Vaccinated	221,998	76	34	56	25	38	17	18	8	—	—	20	9
Controls†	725,173	439	61	391	54	330	46	61	8	11	2	48	6
Grade 2 not inoculated	123,605	66	53	54	44	43	35	11	9	—	—	12	10
Incomplete vaccinations	9,904	4	40	4	40	4	40	—	—	—	—	—	—

*Includes 8,577 children who received one or two injections of placebo.
†First- and third-grade total population.

Source: Adapted from T. Francis, Jr. (1955), Tables 2 and 3.

in antigenic potency did just as well as those judged superior. Another surprise was the rather high frequency with which apparently typical cases of paralytic polio were not confirmed by laboratory test. Nonetheless, there were no surprises of a character to cast serious doubt on the main conclusion. The favorable reaction of those most expert in research on polio was expressed soon after the results were reported. By carrying out this kind of study before introducing the vaccine, it was noted, we had facts about the Salk vaccine that we still lack about the typhoid vaccine and about the tuberculosis vaccine, after many decades of use.

EPILOGUE

It would be pleasant to report an unblemished record of success for the Salk vaccine following so expert and successful an appraisal of its effectiveness, but it is more realistic to recognize that such success is but one step in the continuing development of public health science. The Salk vaccine, although a notable triumph in the battle against disease, was relatively crude and, in many ways, not a wholly satisfactory product and it was soon replaced with better ones.

The report of the field trial was followed by widespread release of the vaccine for general use, and it was discovered very quickly that a few of these lots actually caused serious cases of polio. Distribution of the vaccine was then halted while the process was reevaluated. Distribution was reinitiated a few months later, but the momentum of acceptance had been broken, and the prompt disappearance of polio that researchers hoped for did not come about. Meanwhile, research on a more highly purified killed-virus vaccine and on several live-virus vaccines progressed, and within a few years the Salk vaccine was displaced in the United States, but not in Sweden, by live-virus vaccines.

The long-range historical test of the Salk vaccine, in consequence, has never been carried out. We do not know with certainty whether or not that vaccine could have accomplished the relatively complete elimination of polio that has now been achieved. Nonetheless, this does not diminish the importance of its role in providing the first heartening success in the attack on this disease, a role to which careful and statistically informed experimental design contributed greatly.

PROBLEMS

1. Using Figure 1 as an example, explain why a control group is needed in experiments where the effectiveness of a drug or vaccine is to be determined.

2. Explain the need for control groups by criticizing the following statement: "A study on the benefits of vitamin C showed that 90% of the people suffering from a cold who take vitamin C get over their cold within a week."

3. Explain the difference between the observed control approach and the placebo control approach. Which one would you prefer, and why?

4. Why is it important to have a double-blind experiment?

5. If double-blind experiments provide the only satisfactory way to avoid observer bias, why aren't they used all the time?

6. If only volunteers are used in an experiment, instead of a random sample of individuals, will the results of the experiment be of any value? What can you say about the results?

7. Why did the polio epidemics seem to hit hardest those who were better off hygienically?

8. Why was a *large-scale* field trial needed to get convincing evidence of the Salk vaccine's effectiveness?

9. Refer to Figure 1. In which year did the highest polio incidence occur? The lowest? The largest increase? The smallest increase? Give the approximate values of these incidences and increases.

10. Refer to Figure 1. Comment on the use of the *number of cases.* Can you suggest a different indicator of the spread of poliomyelitis in the United States during 1930–1956? When are the two indicators equivalent? (Hint: Refer to Table 1.)

REFERENCES

K. Alexander Brownlee. 1955. "Statistics of the 1954 Polio Vaccine Trials." *Journal of the American Statistical Association* 50(272): 1,005–1,013.

Thomas Francis, Jr., et al. 1955. "An Evaluation of the 1954 Poliomyelitis Vaccine Trials—Summary Report." *American Journal of Public Health* 45(5): 1–63.

Paul Meier. 1957. "Safety Testing of Poliomyelitis Vaccine." *Science* 125(3,257): 1,067–1,071.

D. D. Rutstein. 1957. "How Good Is Polio Vaccine?" *Atlantic Monthly* 199:48–51.

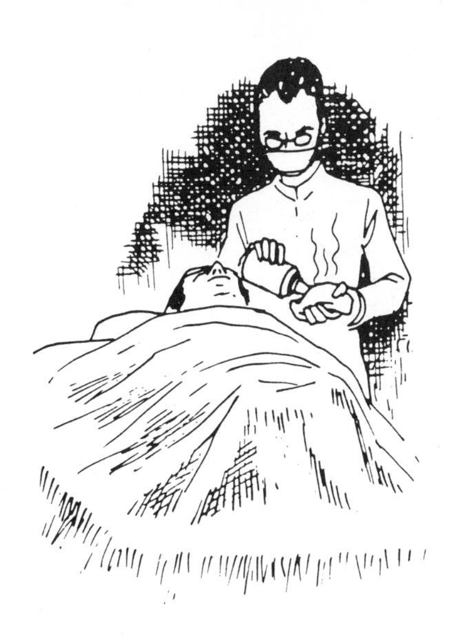

Safety of Anesthetics

Lincoln E. Moses *Stanford University*
Frederick Mosteller *Harvard University*

In 1958, U.S. hospitals began using a new anesthetic called halothane. It soon became widely accepted for its many desirable properties. Unlike some of the more commonly used anesthetics, it could not catch fire, so the potential hazard of a fire or explosion did not have to be a concern during surgical operations. Patients found it less disagreeable and recovered from anesthesia more quickly and with less severe aftereffects. Extensive laboratory research and trials on animals and humans in surgery had encouraged belief in its safety. So there were good reasons for halothane to come rapidly into widespread use. By 1962, surgeons used halothane in half of their operations.

After a few years, however, halothane came under suspicion as accounts appeared in the medical literature of some strikingly unusual—but strikingly similar—deaths of patients who had recently had this anesthetic. A few patients recovering from surgery suddenly took a turn for the worse, ran fevers, and died; subsequent autopsies revealed massive fatal changes in their livers. Even though there were only a few such reports, it was natural to ask, Do these incidents mean that halothane is a poison dangerous to people's livers? Should the use of halothane as an anesthetic in surgery therefore be discontinued? Some compounds chemically similar to halothane were already known to cause liver damage, so the question arose with all the more force. A national committee was appointed to assemble and examine evidence that might help to answer these questions.

Much information was needed. First, it was possible that the liver changes found with halothane might occur equally often with other anesthetics but that physicians rarely reported such cases because the old anesthetics attract less attention. Second, and this is a key point, whether or not halothane had a special adverse effect on the liver, there was the possibility that its other advantages might result in a lower overall postoperative death rate. Frequently things that are easy to use actually work better (this is true of sharp knives, fine violins, and easy-to-read instructions), so it might well be that an anesthetic that is easier for the doctor to use might work better and result in somewhat lower death rates during surgery. Information was needed to confirm or refute this possibility.

These questions could not be answered by doing laboratory experiments with mice, for mice might behave differently under the anesthetic than human beings; nor could the questions be answered by looking in books, for the information was not even known; nor was it useful to ask experts for their judgments since different experts had widely different opinions.

THE EVIDENCE

It was necessary to amass a great deal of evidence, so the committee decided to conduct a survey of hospital experience. At that time, halothane had been given to about 10 million people in the United States. Many of these patients had been in a particular group of 34 hospitals that kept good records and whose staffs keenly wanted to answer the very questions we have asked. They cooperated with the committee by sampling their records of surgical operations performed during the years 1960–1964. In addition to recording whether or not the patient died within six weeks of surgery, they gave information on the anesthetic that had been used as well as facts about the surgical procedure and the patient's sex, age, and physical status prior to the operation. Among the 850,000 operations in the study, there were about 17,000 deaths, or a death rate of 2%. The death rates, shown in Table 1, were calculated for each of four major anesthetics and for a fifth group consisting of all other anesthetics. Note

Table 1 Death rates associated with various anesthetics

Halothane	*Pentothal**	*Cyclopropane*	*Ether*	*All Others*
1.7%	1.7%	3.4%	1.9%	3.0%

*Nitrous oxide plus barbiturate.

that these are death rates from all causes, including the patients' diseases, and are not deaths especially resulting from the anesthetic.

NEED FOR ADJUSTMENT

Table 1 suggests that halothane was as safe as any other anesthetic in wide use, but such a suggestion simply cannot be trusted. Medical people know that certain anesthetics (cyclopropane, for example) are used more often in severe and risky operations than some other anesthetics (such as pentothal, which is much less often used in difficult cases). So some or all of the differences between these death rates might be due to a tendency to use one anesthetic in difficult operations and another in easier ones.

Different operations carry very different risks of death. Indeed, death rates on some operations were found to be as low as 0.25% and on others nearly 14%. The change in death rates across the categories of patients' physical status was even more dramatic, ranging from 0.25%in the most favorable physical status category to over 30% in the least favorable. Age mattered a great deal: the most favorable 10-year age group (10 to 19) carried a death rate of less than 0.50%, while the least favorable age group (over 90) had a death rate of 26%. Another factor was sex, with women about two-thirds as likely to die. Because the factors of age, type of operation, physical status, and sex were so very important in determining the death rate, it was clear that even a relatively small preponderance of unfavorable patients in the group receiving a particular anesthetic could raise its death rate quite substantially. Thus the differences in Table 1 *might* be due largely, and possibly even entirely, to discrepancies in the kinds of patients and types of operations associated with the various anesthetics. Certainly it was necessary to somehow adjust, to equalize for, type of operation, sex, physical status, and age before trying to determine the relative safety of the anesthetics.

If these data had been obtained from a properly planned experiment, the investigators might have arranged to collect the data so that patients receiving the various anesthetics were comparable in age, type of operation, sex, and physical status. As it was, the data were collected by reviewing old records, so substantial differences in the kinds of patients receiving the various anesthetics were to be expected and, indeed, were found. For example, cyclopropane was given two or three times more often than halothane to patients with bad physical status and substantially more often than halothane to patients over 60 years of age. There were many such peculiarities, and they were bound to affect the death rates of Table 1.

CARRYING OUT ADJUSTMENTS

The task was to purge the effects of these interfering variables from the death rates corresponding to the five anesthetics. Fortunately, it was possible to do this. We said "fortunately" because it could have been impossible; for example, if each anesthetic had been applied to a special set of operations, different from those in which other anesthetics were used, then differences found in the death rates could perfectly well have come from differences in the operations performed and there would be no way to disentangle operation from anesthetic and settle the question. But in the data of this study, there was much overlapping among the anesthetics in the categories of age, kind of operation, sex, and patient's physical status, so that statistical adjustment, or equalization, was possible. Analysis using such adjustment was undertaken in a variety of ways because the complexity of the problem made several different approaches reasonable and no one approach alone could be relied upon. There was close agreement in the findings regardless of the method used.

A summary of the results is shown in Table 2. Note that halothane and pentothal after adjustment have higher death rates than their unadjusted rates in Table 1. Also, cyclopropane and "all others," the two with the highest rates in Table 1, now have lower death rates. The effects of these adjustments are quite important: halothane, instead of appearing to be twice as safe as cyclopropane—the message in Table 1—now appears to be safer by only about one-fifth.

Figure 1 shows the adjustment effect graphically.

A Special, Simple Case.[1] To understand the nature of the adjustments, it is helpful to look at a very special case with fictitious data. Suppose that halothane and cyclopropane are to be compared, that we are to adjust only for physical status, and that we classify all patients as in either good or poor physical status. The fictitious data might look like those shown in Table 3. Thus halothane was given to patients in good status four out of five times (in this fictitious case), but cyclopropane, just the other way around, was given to patients in good physical status one out of five times. The two anesthetics have exactly the same death rates for each physical status classification separately, but the very different proportions of status for the two make the overall death rate for cyclopropane almost twice that of halothane.

Suppose that we decide to adjust by computing what the overall death rates *would* be if the two anesthetics were given to a population of patients in which 70% were of good status and 30% poor. The death rates for each status group remain the same. The computations would be as displayed in Table 4. For each anesthetic the overall death rate is now 1.6% with these fictitious data.

In practice, the actual adjustments are far more complicated, partly because fine breakdowns of the data leave many cells with no entries at all and many with small entries.

Table 2 Adjusted death rates for various anesthetics (adjusted for age, type of operation, sex, and physical status)

Halothane	*Pentothal*	*Cyclopropane*	*Ether*	*All Others*
2.1%	2.0%	2.6%	2.0%	2.5%

[1]This section may be skipped without disturbing continuity.

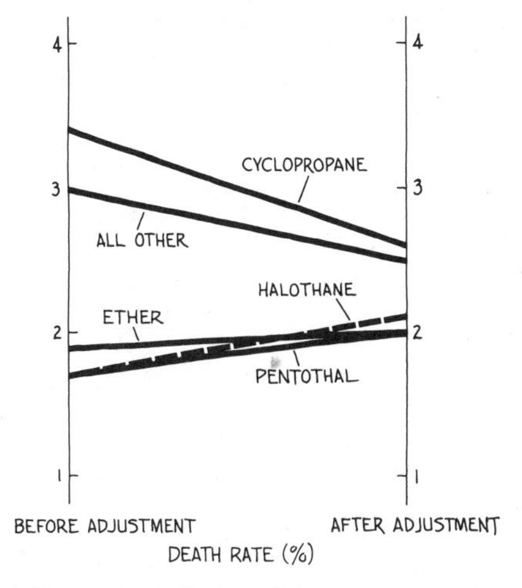

Figure 1 *Effects of adjusting death rates. Plotted points are from Tables 1 and 2.*

Table 3 Fictitious data

	Number of Patients	*Number of Deaths*	*Death Rate*
Halothane			
Poor physical status	200	6	3.0%
Good physical status	800	8	1.0%
All patients	1,000	14	1.4%
Cyclopropane			
Poor physical status	800	24	3.0%
Good physical status	200	2	1.0%
All patients	1,000	26	2.6%

Table 4 Adjusted death rates (fictitious data)

	Number of Patients	*Number of Deaths*	*Death Rate*
Halothane			
Poor physical status	300	9	3.0%
Good physical status	700	7	1.0%
All patients	1,000	16	1.6%
Cyclopropane			
Poor physical status	300	9	3.0%
Good physical status	700	7	1.0%
All patients	1,000	16	1.6%

Consistency over Hospitals. Even after the adjustments, there was still considerable variability from hospital to hospital. Our confidence in the validity of the statistically adjusted figures in Table 2 is affected by how consistent the death rates were from hospital to hospital. After all, if there were many hospitals where cyclopropane had a lower adjusted death rate than did halothane, even though halothane was lower "on the average," we might feel uncertain that halothane was really safer just because of the difference in adjusted death rates. Some fluctuation from hospital to hospital is to be expected, of course, because of chance factors; absolute consistency is not to be expected.

The question is: Were the comparisons between the adjusted anesthetic death rates consistent enough over hospitals to warrant taking seriously the apparent differences shown in Table 2? Rather complicated statistical techniques were necessary to study this question, but the conclusions were clear: halothane and pentothal both had adjusted death rates definitely lower than the adjusted death rates for cyclopropane and "all others," and those lower adjusted death rates were real in the sense that they could not be explained by chance fluctuations. The differences were sufficiently consistent to be believed. On the other hand, ether, with the same adjusted death rate as halothane and pentothal, did not have a consistent pattern of comparison. Therefore, ether cannot be reliably compared with the others; we cannot tell from the evidence obtained whether it may be somewhat safer than it appears here—or somewhat less safe. Possibly it is as safe as, or safer than, halothane; possibly it is no safer than "all others." The findings about ether are indefinite because there were fewer administrations of ether and these were concentrated in a few hospitals; hence there are fewer data to go on.

A SUMMING UP SO FAR

What can we conclude so far from the findings of this study? First, and most important, is the surprising result that halothane, which was suspect at the beginning of the study, emerged as a definitely safe and probably superior anesthetic agent. Second, we see that a careful statistical study of 850,000 operations enabled the medical profession to answer questions more firmly than had previously been possible, despite the much greater "experience" of 10 million administrations of halothane and other tens of millions of administrations of the other anesthetics.

HOSPITAL DIFFERENCES AGAIN

A brand-new question emerged with the observation that the 34 hospitals had very different overall postoperative death rates. These ranged from around 0.25% to around 6.5%. This seemed to mean that the likelihood of dying within six weeks after surgery could be more than 25 times as great in one hospital as in another. Just as before, however, there were strong reasons to approach this startling information skeptically. Some of the hospitals in the study did not undertake difficult operations, such as open-heart surgery, while others had quite large loads in such categories. This kind of difference alone would cause differences in hospital death rates. Furthermore, the age distribution might be different, and perhaps importantly so, from one hospital to another. Indeed, one was a children's hospital, another a veteran's hospital. Some hospitals might more frequently accept surgical patients having poor physical status. So the same interfering variables as before surely affected the differences in hospital death rates. If adjustment was made for them, would the great differences in hospital death rates vanish? Be much reduced? Remain the same? Or, as is conceivable, actually increase?

Adjustment procedures were applied, and the result was that the high-death-rate hospitals, after adjustment, moved down toward a 2% overall death rate, and the low-death-rate hospitals, after adjustment, moved up toward a 2% overall death rate. The adjusted hospital death rates no longer ranged from 0.25% to over 6.5%; instead, the largest of these adjusted death rates was only about three times as great as the smallest. Now, almost any group of 34 such rates will exhibit some variability, and the ratio of the largest to the smallest must be a number larger than 1. Even if a single hospital were measured over several different periods, the rates would fluctuate from chance alone. The fluctuation in rate would be considerable because the death rate itself is basically low and one death more or less makes a difference in the observed rate for the period. The fact that this ratio turned out to be 3 in these data does not, in itself, indicate clearly that there were real, unexplained hospital differences.

Careful statistical study showed that there are probably some real differences from hospital to hospital in postoperative death rates and that these differences cannot be explained wholly by the hospitals' patient populations in terms of age, sex, physical status, and surgical procedure. Statistical theory showed that the ratio of highest to lowest adjusted rate should be about 1.5 if the hospitals were identical in operative death rate after adjustment (so that the adjusted rates differ only by chance fluctuation).

The position, then, is that we began with the large ratio of about 25 for unadjusted rates, cut the ratio way down to 3 for adjusted rates, and then compared the 3 with 1.5 as a theoretical ratio. Because the adjustments can hardly have been perfect and because there undoubtedly were unadjusted factors that differed among the hospitals, we conclude that the adjusted hospital death rates are indeed close together. Thus what on the basis of the unadjusted hospital death rates looked like a shocking public health problem proved, after statistical investigation, to be quite something else. The apparent problem was mainly, though perhaps not entirely, a dramatic manifestation of differences not in the quality of surgical care but in the difficulty of the surgical cases handled in the various hospitals.

FOLLOW-UP STUDIES

Concern about hospital differences prompted further extensive studies of hospitals in the United States to determine whether the differences in surgical death rate among hospitals are as large as the halothane study suggested.[2] These follow-up studies were designed to gain further information about postsurgical complication rates and to begin exploring the reasons for hospital differences if such differences were confirmed. (See Flood and Scott, 1987.)

The studies have confirmed the results of the halothane study and have added further detailed estimates on the sizes of the differences and on the kinds of surgical procedures that vary most. Furthermore, by carefully adjusting the hospital rates for the characteristics of the patients served by the hospital, and then correlating hospital characteristics with measures of hospital performance, some interesting hypotheses have been developed for further study. For example, the results suggest that a very important determinant of the quality of surgical outcome is the care and power exercised by the hospital physicians in approving surgeons for hospital privileges in their hospital.

LIVER DAMAGE

Let us return to the question of liver damage. The total deaths from this source were few, and the lack of autopsies for nearly half of the deaths made firm conclusions impossible. Since the study was made, however, an anesthesiologist, who had often been exposed to halothane while administering it to patients, has been discovered to be sensitive to halothane; he exhibits symptoms of liver malfunction from breathing it.

Are occasional individuals sensitive to other anesthetics? How many people develop such sensitivity? These are hard questions to answer, and they are especially difficult to study because of the rarity of the occurrences.

CONTRIBUTIONS OF STATISTICS

What were the main contributions of statistics in the program? First was the basic concept of a death-rate study. This needed to be carried out so that the safety of anesthetics could be seen in light of total surgical experience, not just in deaths from a single rare cause. Second, though we have not discussed it, the study used a special statistical technique of sampling records, designed to save money and to produce a high quality of information. Third, special statistical adjustments had to be created to appraise the results when so many important variables—age, type of operation, sex, patient's physical status, and so on—were uncontrolled. Fourth, as a result, the original premise of the study was not sustained, but a new result emerged: halothane seemed safer than cyclopropane. The study produced a nonfact as well: the comparative merit of ether is uncertain. We cannot tell if its associated death rate is higher, lower, or nearly the same as halothane, though the indication is that it is nearly the

[2]Staff of Stanford Center for Health Care Research, "Comparison of Hospitals with Regard to Outcomes of Surgery," *Health Services,* Summer 1976, pp. 112–127.

same. Fifth, new evidence on hospital differences showed that the initial, wide variation in death rates, when suitably adjusted for patient populations, left only small unexplained differences in rates. These remaining differences led the medical profession to begin, in 1971, a study of possible causes of hospital-to-hospital differences in postoperative death rates, in the hope of discovering ways to improve postoperative care.

AFTERWORD

In 1988, halothane, ether, and cyclopropane are seldom used. They have been replaced by newer anesthetics.

The nation has a fresh concern about the quality of hospital care because the system of reimbursements introduced in the 1980s encourages hospitals to discharge patients sooner or often to treat them through ambulatory centers rather than admit them to a hospital at all. The resulting reductions in numbers of admissions and in length of hospital stay have led to fears of lowered quality, but we have little proof either way. Methods akin to those of this essay are being applied to clarify matters.

Usually attempts to measure the quality of hospital care are based on mortality figures adjusted for the severity of illness and other characteristics of the patients. Many researchers believe, however, that such adjusted death rates do not properly measure the quality of care in a developed country. Instead they would review other outcomes, such as complications and readmissions.

Although halothane itself has been replaced, the National Halothane Study has been important in two other ways: first, it encouraged statisticians to develop methods that now are in widespread use, not only in medicine and health but in other areas as well; second, the study's results are still the only source of information concerning certain operative death rates adjusted for various risk factors.

PROBLEMS

1. Explain how it was possible for halothane to be as safe as other anesthetics despite the evidence from autopsies that it was associated with liver damage.
2. What is meant by equalization or adjustment?
3. Explain the discrepancy between the safety of halothane indicated in Tables 1 and 2.
4. How can you explain the unadjusted differences of the overall postoperative death rates in the 34 hospitals in the study?
5. What were the principal conclusions of the study?
6. What statistical tools were used in the study?
7. Explain why the results about ether were uncertain.

8. Is halothane a cause of liver damage? Explain.
9. Refer to Figure 1. For which anesthetic is the adjustment most drastic? Check your answer by comparing Table 2 with Table 1.
10. In Table 4 explain how the number of deaths in each status group was calculated.
11. Consider a study of the effect of a high cholesterol diet on mortality from coronary heart disease (CHD). The mortality rates from CHD in two groups are compared—one has a high cholesterol diet and the other has an average diet. What are some of the factors that one needs to adjust for in the two groups before comparing the mortality rates?
12. Why did the statistical study described in this article concentrate on 850,000 operations only although there were over 10 million operations in which halothane was administered and tens of millions of operations where other anesthetics were used?

REFERENCES

John P. Bunker, William H. Forrest, Jr., Frederick Mosteller, and Leroy D. Vandam. 1969. *The National Halothane Study,* Report of the Subcommittee on the National Halothane Study of the Committee on Anesthesia, Division of Medical Sciences, National Academy of Sciences—National Research Council. Washington, D.C.: U.S. Government Printing Office.

Ann Barry Flood and W. Richard Scott with Byron W. Brown, Jr. and others. 1987. *Hospital Structure and Performance.* Baltimore, MD: Johns Hopkins University Press.

The Metro Firm Trials and Ongoing Patient Randomization

Duncan Neuhauser *Case Western Reserve University*

Many large hospitals in Europe and North America have teams of physicians who care for groups of patients. New patients are assigned to a team in rotation. The first patient goes to team A, the second to team B, and so on, over and over. This is usually done to equalize the work load of the teams. The teams can include physicians, nurses, and other staff. They are responsible for outpatients, inpatients, or both. Although these arrangements have been in existence since the turn of the century, no one has ever taken advantage of such an arrangement to accurately evaluate the effects of changes in care and done this on an ongoing basis.

The 700-bed Cleveland Metropolitan General hospital (Metro) has four such teams, called "firms," of general internal medicine physicians who work with nurses and other allied health personnel. Each firm has 18 physicians, 28 inpatient beds, and an outpatient clinic. From 1976 to 1981 all new patients were assigned in rotation. Starting in 1981, patients were assigned by a computer program that generates the numbers 1, 2, 3, and 4 continually in random sequence. This random assignment provides excellent assurance that the four groups of patients are nearly identical. If the assigned firm's inpatient unit is full, the patient is "boarded" on another unit until a bed is free and then the patient is transferred.

In every hospital, hundreds of changes in the organization and delivery of medical care are made in the belief that such changes provide better care and/or more efficient care. Only a small fraction of these changes are carefully evaluated to show whether this belief is correct.

Because patients are different, evaluation of medical care is particularly difficult. One has to be sure that observed changes in the results (for example, better care or lower cost) are due to the care itself rather than to patient differences. The history of medicine is marked by unevaluated, probably harmful, treatments, from blood-sucking leeches to surgical lobotomies for mental illness, which have been applied to tens of thousands of patients. There is a continuing need for careful evaluation so that such useless and harmful procedures happen less often in the future.

Since 1981 at Metro, the four firms and their similar patient populations have been used to evaluate an ongoing series of changes. In these trials, a change is put in place in two randomly chosen firms, while the other two firms are left as they are to serve as controls. To increase the assurance of comparability, since 1983 all new resident internal medicine physicians have been randomly assigned to one of the four firms. These residents remain with their firm for the three years of their training.

USING THE FIRMS

A series of trials has been carried out under the direction of David Cohen, M.D., using the Metro firms. These studies asked such questions as: Does feedback to resident physicians about the cost of the laboratory tests they order reduce their use of tests? Yes. Do reminders to carry out appropriate preventive procedures for ambulatory patients increase their use? Yes. Does better organization of outpatient care reduce the cost of inpatient care? Yes.

We describe two trials in more detail. One focused on better quality of care and the other on lowering costs. Ordinarily at Metro, intravenous or IV therapy (feeding patients fluids by tube into an arm vein) was managed by the regular nurses and physicians. Some physicians thought having a specialized IV team would reduce the rate of infections and complications (phlebitis). The IV teams were put in place in two firm inpatient units, and the percentage of patients on IVs with infections was compared with firms managing IVs in the usual way. The patients of the specialized IV teams did have significantly fewer infections. The teams had 15% compared to the control groups' 32%. Very serious complications were reduced from 2.1% to 0.2% by these specialized teams. As a result of this evidence, the specialized teams were used in all four firms. Later, during a cost-control drive, it was proposed that the IV teams be eliminated.

However, by showing the evidence of their efficacy, the IV team members kept their jobs, and the patients continue to benefit.

The hospital now has a new computerized information system, which includes information about patients, their physicians, the firm they are assigned to, laboratory tests, and drugs used. In the second study, Charles Hershey, M.D., and colleagues used this firm-and-information system to enlighten physicians about drug usage. They sent a computerized message to two firms suggesting that physicians substitute less costly drugs for an expensive, but otherwise similar, drug and indicated how much money each doctor could save by doing so. Then, using the same information system, they measured the effect of this message. They found a significant substitution effect for the experimental firms (see Table 1). What was of particular interest about this trial was the very low cost of carrying it out. Because of the ongoing randomization and the computerized information system, the additional cost for this study was less than a thousand dollars. The largest cost was the time Dr. Hershey took to write the report for publication.

GENERALITY OF FINDINGS

It is legitimate to ask whether Metro and its patients are representative of all large North American hospitals. What works at Metro may not work at other hospitals. Even so, just improving care at this one hospital is a good thing to do. Other hospitals could develop similar systems of ongoing patient randomization to their own benefit.

If similar studies were carried out in several such hospitals, we could begin to learn how representative Metro really is. One such program has been started at University Hospitals of Cleveland, where all new ambulatory patients cared for by the general medicine clinic are now randomly assigned to one of two teams of physicians by the last digit of the patient's Social Security number. Even numbers go to one team, odd numbers to the other. The clinic director, Victoria Cargill, M.D., is now completing a trial to see if it is more effective

Table 1 Mean charge for drugs used before and after computerized feedback of information to physicians; comparing experimental and control firms 1984–1985

	Before Study Started			*At End of Study*		
Firms	*Mean Drug Cost**		*Standard Deviation***	*Mean Drug Cost*		*Standard Deviation*
Experimental	$8.44	±	1.28	$8.22	±	1.01
Control	$8.56	±	1.89	$8.79	±	0.83

* These costs are based on the use of three common drugs by 48 physicians. The difference between the control and experimental firms at the end of the study is significant and very unlikely to have occurred by chance (Hershey, Porter, Breslau, and Cohen, 1986). Although these differences may look small per prescription, this difference applied to all firms at Metro would result in a savings of $6,500 per month.

** A standard deviation is a measure of spread or variation. In many distributions about two-thirds of the data fall within a distance of 1 standard deviation from the mean and about 95% fall within 2 standard deviations.

to have nurse practitioners rather than the resident physicians screen for colon cancer. Dr. Cargill found that patients seen by the nurse were much more likely to send their take-home test results back to the hospital for evaluation and, if need be, follow-up evaluation and treatment.

The earliest known randomization of patients for the purposes of evaluating treatment was published in 1931. (Amberson et al. divided 24 tubercular patients into two comparable groups of 12 each, then matched them in pairs. By a flip of a coin one group of 12 was chosen to be treated with sanocrysin and the other became a control group.) But it was only in the 1950s that randomization began to be applied widely. Single randomized trials can be very expensive—some have cost more than a hundred million dollars. If such studies lead to more rapid diffusion of beneficial treatment or discourage a costly, worthless treatment, however, the worldwide benefits would be well worth such large costs. We may hope that making good evaluation less costly, less difficult, and more routine will lead to more of it. Efficient, repeated evaluation of changes that lead to better care is the special contribution of ongoing patient randomization. It is a general idea that could be applied outside the health care field.

The world has no large set of organizations prepared to carry out medical evaluations through experimentation at a moment's notice in hospitals. By and large each experiment has to be organized separately at considerable expense, and because it is organized especially for the occasion, it does not exactly represent the medical practice that is in place in the institutions. Therefore it is difficult and expensive to get evaluations of technologies done in the medical and health fields.

On the other hand, the same thing used to be true of the field of sample surveys. Every sample survey had to be set up from scratch because no national organization was in place, and so surveys were especially expensive because of organization and start–up costs. Now there are many national organizations that can carry out surveys quickly and economically because the organization and the expertise are all there and ready to go. We need such an arrangement for the evaluation of medical technologies. Such evaluations must be done primarily through *experiments,* which are much harder to execute than sample surveys. Thus it is important to have a collection of organizations (hospitals with firms) ready to carry out evaluations through experimentation with medical technology.

PROBLEMS

1. New patients are randomly assigned to the firms in order to provide the best assurance that the patient groups cared for by each firm are similar. If these patients need care, from then on they go back to the same firm. As time passes do you think the patient groups seen by the firms will continue to be similar? Explain your answer.

2. Do you think you could apply this concept to a large high school or college? How would you do it?

3. Can you pick some other organizations where this concept could be applied? What organizations? How would you do it?

4. In Reykjavik, the capital of Iceland, there are three government hospitals. On the first day, all emergency patients go to hospital A, on the second day, all go to hospital B, on the third day all go to hospital C, then the cycle is repeated continuously. The Ministry of Health has asked you to advise them about the quality and cost of care in their hospitals. What would you do? (You might measure quality of care by death rates, length of sickness, or time lost from work.)

5. You are responsible for stopping drug smugglers at the Miami International Airport where thousands of passengers arrive every day from many countries. You are in charge of 200 customs inspectors. Every passenger passes through one of your 20 customs checkout stations. You have lots of ideas about different ways to find drugs, but you are not sure they will really work. (For example, you could have every large suitcase opened and inspected. But this would take longer than opening just suspicious suitcases, and passengers get upset when they have to wait in line too long.) Explain how you might find out if your ideas are good ones.

6. You are assistant to the president of an international company that owns 2,000 retail ice cream stores. The vice president for marketing wants to offer cinnamon-flavored cones with red spiral stripes in addition to the usual brown-colored cones. The president has asked you to develop a plan to evaluate this and other new ideas. What would you do?

REFERENCES

J. B. Amberson, Jr., B. T. McMahon, and M. Pinner. 1931. "A Clinical Trial of Sanocrysin in Pulmonary Tuberculosis." *American Review of Tuberculosis* 24:401–435.

V. Cargill, D. Cohen, K. Kroenke, and D. Neuhauser. 1986. "Ongoing Patient Randomization: An Innovation in Medical Care Research." *Health Services Research* 21:663–678.

D. I. Cohen, D. Breslau, et al. 1986. "The Cost Implications of Academic Group Practice: A Randomized Controlled Trial." *New England Journal of Medicine* 314:1553–1557.

D. I. Cohen, P. Jones, B. Littenberg, and D. Neuhauser. 1982. "Does Cost Information Availability Reduce Physician Test Usage? A Randomized Clinical Trial with Unexpected Findings." *Medical Care* 20:286–292.

D. I. Cohen, B. Littenberg, C. Wetzel, and D. Neuhauser. 1982. "Improving Physician Compliance with Preventive Medicine Guidelines." *Medical Care* 20:1040–1045.

D. I. Cohen and D. Neuhauser. 1985. "The Metro Firm Trials: An Innovative Approach to Ongoing Randomized Clinical Trials." In Institute of Medicine, *Assessing Medical Technologies.* Washington, D.C.: National Academy Press, pp. 529–534.

C. O. Hershey, D. K. Porter, D. Breslau, and D. I. Cohen. 1986. "Influence of Simple Computerized Feedback on Prescription Charges in an Ambulatory Clinic." *Medical Care* 24:472–481.

C. O. Hershey, J. W. Tomford, et al. 1984. "The Natural History of Intravenous Catheter-Associated Phlebitis." *Archives of Internal Medicine* 144:1373–1375.

J. W. Tomford, et al. 1984. "Intravenous Therapy Team and Peripheral Venous Catheter-Associated Complications: A Prospective Controlled Study." *Archives of Internal Medicine* 144:1191–1194.

A Health Insurance Experiment

Joseph P. Newhouse *The RAND Corporation*

Most Americans have health insurance that helps them pay hospital and doctor bills. The key word is *helps*—only a very unusual insurance policy will pay the whole cost for all types of services. Most insurance policies require "cost sharing" by the patient through such mechanisms as deductibles and coinsurance. (A *deductible* of $200 per person per year requires the patient to pay the first $200 of medical bills in a year; *coinsurance* of 20% requires the person to pay 20% of every bill, above any deductible.) In some countries, however, public insurance may pay for all medical bills with no cost sharing.

Someone must decide how large deductibles and coinsurance rates will be—or whether there will be any cost sharing at all. The person deciding may be an employer or a union negotiator. In the Medicare and Medicaid programs, the decision makers are the Congress and the president. How should these people decide? What information should they want to know? It seems they should want to know how much more health is bought by a more generous, and hence more expensive, insurance policy.

This issue has caused controversy for decades. Some have argued as follows: If people must pay for medical care, they will postpone or entirely forego needed services, and their health will be jeopardized. Those arguing this way usually conclude that medical care should be free. They tend to minimize the additional cost of care, taking the position that people do not enjoy going to the doctor and only go in times of real need.

Others have argued just the contrary: If medical care is free, people will abuse it and make many unnecessary doctor visits that do not affect their health. As a result, this group alleges that free medical care will be very costly, and they usually conclude that people should pay for most of their medical care, with insurance covering only very large medical bills.

Both sides have argued with more passion than evidence. To get more information about the issues, the federal government in 1974 set up a large experiment, the Health Insurance Experiment. People participating in this experiment were randomly assigned to different insurance plans that required them to pay differing amounts for their medical care. Both their health and the amount of medical care they used were monitored for several years. In its size and goals, this experiment in health policy was unprecedented.

Let's look at some examples of how statistics was used to design the experiment and to analyze the data collected in order to obtain as much information as possible for the money spent on the experiment.

THE STATE OF KNOWLEDGE AT THE BEGINNING OF THE EXPERIMENT

Why did we know so little in 1974? After all, health insurance was hardly novel. By 1974, when the experiment began, private insurance or the government had paid 80% to 90% of all hospital bills for two decades and insurance had paid over half of all doctor bills. One might have thought that the debate could be largely settled without an experiment simply by studying the actions of people with existing insurance policies.

Indeed, insurance companies could quote premiums on different insurance policies, and we could use those premiums to estimate the total cost to the person (including any cost sharing) of alternative insurance plans. But trying to shed light on cost and health effects of insurance by using data on existing insurance policies faces three difficulties:

1. People choose insurance policies with an eye to how much they would likely spend on medical care. Those who anticipate large expenditures have a clear incentive to seek out policies that cover their medical bills rather extensively. By contrast, those fortunate individuals who do not expect to see a doctor at all see little benefit in paying a large premium for nearly full coverage and so tend to opt for a policy with a small premium and large cost sharing. As

a result, if we try to infer the use of those not now on generous plans from the experience of those who have generous plans, or vice versa, we might considerably overstate the effects of insurance on use and cost.

2. Even if we could reliably assess the use of services (and thus the costs) under alternative insurance plans, there were no data on the effects of alternative plans on health. Yet this was the crux of the issue for many people who agreed that better insurance led people to seek help more often, but disagreed as to whether this help was necessary or unnecessary. Even if we obtained information on individuals' present health, insurance usually changes when people change jobs or retire; thus we could not infer whether present health was related to the present or the past insurance policy.

3. Finally, some insurance arrangements simply had not been tried. For example, some advocates of cost sharing thought cost sharing should be scaled down for the poor. But the poor typically either had no insurance or were covered in full by their state's Medicaid program. How the poor might respond to a plan that had them pay, say, up to the first 5% of their income for medical care could not be learned from existing data.

THE DESIGN OF A SOCIAL EXPERIMENT

Around 7,700 people who lived in six different areas of the country participated in the Health Insurance Experiment. They did not use their own health insurance (if any) for either three or five years and instead used an experimental health insurance plan. At the end of the three or five years, they returned to their usual health insurance.

Some participants received free medical care; others paid 25%, 50%, or 95% of their medical bills, up to a maximum of $1,000 out-of-pocket per family per year. This maximum was reduced for poorer families; specifically, it was never more than 5%, 10%, or 15% of a family's income, depending on the plan to which they were assigned.

This sketch of the Health Insurance Experiment has already encompassed a multitude of decisions on its design. Some of the issues decided include: (1) How many people will participate, and how many will be assigned to each plan? (2) How many sites will be used? (3) Exactly which people will be assigned to which plans? (4) How long will the experiment last? Statistical thinking helped the planners reach those decisions.

The Number of People in the Experiment

Because the objective of any experiment is to produce information, the experiment's designer seeks to maximize the information yielded for the monies spent and to inform those paying for it about how much more information they can buy for more money. One way to measure the amount of information is the precision of the results, or how likely it is that the results of the experiment (for example, in this experiment, the number of doctor visits per person per year on an experimental plan) approximate the true state of the world (for example, how many times per person per year people would seek medical help if everyone had that insurance plan). The likelihood of the experimental results being close to the true number can always be increased by enrolling more people in the experiment. But we get diminishing returns from enrolling more people; for example, increasing the sample from 50 persons to 100 persons gains more than increasing it by another 50 persons to 150 persons. In fact, the precision of any given sample is generally proportional to the square root of the sample size.

Although larger samples give more precise results, they also cost more. Just as a person building a house must decide how much space to buy, the person paying for the experiment has to decide how much precision to buy; the experiment's designer cannot say how much precision should be bought, any more than an architect can say how large a house should be.

Sometimes one sees a statement that implies there is some critical threshold for sample size, and if the sample does not exceed that number, the study is not "scientifically valid." This notion is false; it is like saying a structure must exceed a certain number of square feet or it is not a house.

The Number of Persons per Plan

Now suppose our experiment's budget has been set. The designer must then decide how many people should be assigned to each experimental insurance plan. The first reaction might be to allocate equal numbers of people to each plan, but suppose one plan is much more costly per person enrolled. For example, the insurance plan with free medical care might be more costly than the plan that has persons pay 95% of the bill. Then we do not allocate equal numbers to plans but rather allocate them in accordance with the square root of the reciprocal of the cost ratio. For example, if one insurance plan were four times as costly as another, it would be assigned half as many people as the other plan. We use the square root of the cost ratio because the precision of a sample increases as the square root of the sample size.

The Number of Sites

Cost considerations also affect the appropriate number of sites. Certain costs must be paid just to enter a site; for example, a sampling plan must be drawn up and perhaps space for a site office must be rented. Hence, the more sites, the fewer total persons can be enrolled for any given budget. Our first thought might be to maximize the number of persons enrolled by choosing all persons to be in a single site, but we might have second thoughts when deciding which site to choose. If all sites were alike, one site would be optimal. But because sites differ (for example, in how long people stay in the hospital), we gain information by having more sites. In general, the more sites differ, the more sites and the fewer total people one should have. Cost is also relevant; the higher the costs of entering a new site relative to the costs of adding another person at an old site, the fewer sites are optimal.

The Length of the Experiment

The designer with a given budget must also decide whether to enroll fewer persons for a longer period of time or more persons for a shorter period of time. Two considerations favor enrolling more persons for a shorter period of time. A person's behavior tends to repeat itself over time (for example, someone with severe acne may make frequent doctor visits). Hence, two years of observation of one person is not as valuable as one year of observation of two persons because we already have a clue from the first year whether a person is a high user or not. Moreover, enrolling more persons for a shorter period of time makes information from the experiment available sooner.

But, pushed to its logical limit, these arguments imply enrolling as many persons as the budget would permit, each for one day! That seems absurd, and for good reason. Some of the behavior one wants to measure takes time to occur; for example, what happens to a person's health. Thus we want the experiment to be as short as possible, consistent with giving a long enough time period to allow the insurance plan to have its effect.

In choosing the period of time to operate the experiment, the designer must consider that persons enrolled in an experiment may not behave as they would if a program were of indefinite duration. For example, persons with an experimental insurance plan that will end on a certain date may crowd in or postpone certain services depending on what their situation will be after the experiment. Because this behavior would not be observed in an actual (nonexperimental) plan, it can contaminate the results. One strategy that was used in the Health Insurance Experiment to assess the amount of contamination was to enroll persons for different lengths of time, some for three years and some for five years. Any crowding in or postponement of services at the end of the experiment was measured by comparing the last year of a three-year group with the third year of the five-year group that began at the same time. Similarly, any initial surge was measured by enrolling in other sites a five-year group first and then a three-year group two years later.

The Refusals

The designer of any experiment must take into account that (generally) people cannot be forced to participate in an experiment but must instead agree to participate. Moreover, the designer is under an ethical and legal injunction to inform potential participants about the benefits and risks of participating. Although participants must agree and must be informed, the designer controls the benefits, and perhaps the risks, which will influence the potential participants' decisions.

To minimize refusals, the Health Insurance Experiment paid people enough to ensure that they could not lose financially by participating. For example, on one experimental plan the person had to pay 50% of any hospital bill up to a $1,000 maximum in any year. Suppose the person's prior insurance plan would have fully covered any hospital stay. In this case the experiment paid the person $1,000 per year, irrespective of whether the person went to the hospital. If the amount paid had been less than $1,000, persons who expected to go to the hospital with a high degree of certainty (for example, pregnant women) would probably refuse to enroll because they would expect to lose money.

To be sure, the separate payment might itself change behavior—and, if it did, it would be important to know about it because the separate payment would not be part of a "real world" insurance plan. In order to study the payment's effect, the designer can build some variation into the payment, for example, by randomly choosing some families to receive $1,500 rather than $1,000. Such variation was in fact built into the Health Insurance Experiment, and it turned out that the payments had a negligible effect on behavior.

THE ANALYSIS OF THE EXPERIMENTAL DATA

Table 1 shows the use and spending on medical care per person by insurance plan in the Health Insurance Experiment. Clearly how well a person is insured matters a great deal in how many medical care services he or she uses. Persons paying 95% of the bill made 40% fewer doctor visits than those paying nothing. Similarly, hospital usage decreased by 23%, and total spending by 31%.

We want to assess whether the additional services received by the group on the free care plan affected their health. Reliable evidence on this issue had never before been available. The data in Table 2 show that the health of the average person was little affected by the differences in use highlighted in Table 1. (Table 2 groups together the 25%, 50%, and 95% coinsurance plans for ease of interpretation. The outcomes of these three plans were similar on all measures.)

The values shown in Table 2 are three of the many measures that were collected; they include the two measures, blood pressure and vision, where we can be reasonably confident that additional services did lead to an improvement in health. (Certain measures of dental health were also improved.) The overall measure of health, General Health Perceptions, is, however, slightly better

Table 1 Measures of medical care use, per person per year, as a function of insurance*

Percentage of Bill Paid by Person†	*Physician Visits*	*Hospital Admissions per 100 Persons*	*Spending*‡ *(1984 Dollars)*
0	4.55	12.8	777
25	3.33	10.5	630
50	3.03	9.2	583
95	2.73	9.9	534

*Although not shown, the precision of these numbers is quite good. For details, see Manning et al. (1987).
†Percentage was paid up to an annual out-of-pocket maximum, which was $1,000, or 5%, 10%, or 15% of income, whichever was less. The percentage of income was assigned at random.
‡Includes monies paid by insurance and by the person. Excludes spending on dental services and outpatient psychotherapy. Dental services and outpatient psychotherapy exhibit about the same degree of responsiveness to plan; including them would add a little more than one-third to each of the numbers. Most of the additional monies would be for dental services. The numbers in this column come from averaging predicted spending; see Manning et al. (1987).

Table 2 Measures of health outcome as a function of insurance plan*

Type of Insurance Plan	*General Health Perceptions†*	*Diastolic Blood Pressure (mm of Hg)*	*Vision Measure‡ (Snellen Lines)*
Free Care	67.4	77.9	2.42
Cost Sharing	68.0	78.8	2.52

*These values are predicted from a regression equation and hence are somewhat more precise than simple means. For details, see Brook et al. (1983).

†This is a scale that ranges from 0 to 100 (100 being healthiest) and that describes how people rated their own health. It is based on answers to 22 different questions. A 5-point difference in the scale, all else equal, is equivalent to having mild hypertension (high blood pressure); a 7-point difference, all else equal, is equivalent to having diabetes mellitus.

‡An outcome of 2 corresponds to 20/20 corrected vision in the better eye, an outcome of 3 corresponds to 20/25, and an outcome of 4 to 20/30; thus, smaller numbers are preferable.

for the group of individuals who paid something for their medical care (68.0 versus 67.4). More important, though, is the question of how precise this estimate is. Using statistical analysis we can conclude that any true difference in General Health Perceptions between the two groups is likely to be small; specifically, it is likely to be between 1.5 points (favoring cost sharing) and −0.3 point (favoring free care). More specifically, we can say that if a similar experiment was carried out many times, and the same procedure was used to construct this band, 95% of the resulting bands would contain any true difference between the two groups (there may, of course, be no difference). Because the upper limit of the band in this case is 0.3 point, we can be reasonably certain that the free plan is unlikely to be more than 0.3 point better than the cost-sharing plan.

Of course, we have no intuition about what a 0.3-point difference might mean. One way to gain some intuition is to look at all the people in the study who have some disease, for example, hypertension (high blood pressure). If we adjust for observable differences between people who do and do not have hypertension (that is, adjust for the older age of the group with hypertension), the group with hypertension has a value for General Health Perceptions about 5 points less than the value for the group without hypertension. (For an explanation of adjustment see the Moses and Mosteller essay.) Because 5 points is large relative to 0.3 point, one interpretation of these outcomes is that the additional services that the people on the free care plan consumed did not make them noticeably healthier in their eyes. They may have thought themselves a tiny bit healthier and the measurements failed to detect that, but the opposite may also be true.

For people with hypertension and vision problems, we can be relatively sure that beneficial effects occurred. Those effects were concentrated among the group of poor individuals.

Upon reflection it is probably not surprising that these two conditions benefited from additional care. Hypertension and vision problems (and dental disease as well) are widespread relative to other diseases and are relatively inexpensive to diagnose and treat. Moreover, it is plausible that the poor would benefit differentially. A thoughtful reader might ask why, if these effects on blood pressure and vision existed, they were not reflected in a measure such as the General Health Perceptions. One answer is that the proportion of people affected by these problems and the benefits of free care over those of cost sharing for those problems are sufficiently small that, when a measure of health is averaged over the entire population, the effects are not detectable with the sample size in the experiment. Another possibility is that some fraction of the additional services on the free care plan were associated with bad outcomes (for example, reactions to a drug that was prescribed). These bad outcomes may have offset other good outcomes.

THE INFLUENCE OF THE RESULTS

Because the experiment's findings were important and unique, especially those in Table 2, they appear to have had a substantial effect. The results in Table 1 were published in December 1981, and those in Table 2 in December 1983. Between 1982 and 1984 there was a considerable increase in the amount of cost sharing in private health insurance in the United States, especially for hospital services. Rates of hospital admissions declined, as did physician visits. Table 3 shows these changes. Using the results of the experiment, we can infer that reductions in use in response to greater cost sharing in private insurance probably had little or no adverse effect on health—because most people with private insurance are not poor.

Of course, we cannot attribute all the changes shown in Table 3 to the information provided by this experiment; indeed, we can never know how much of a role the experiment played in bringing them about. But the experiment's results are well known to consultants who advise companies on their health

Table 3 Change between 1982 and 1984 in insurance and medical care utilization

	1982	*1984*
Percent of private insurance plans with a deductible for hospital services	30	63
Percent of private insurance plans with a deductible of $200 per person per year or more	4	21
Hospital discharges per 100 persons under 65*	11.8	10.7
Physician visit rate per person under 65	4.9	4.7

*Excludes deliveries.

Source: First two rows: Hewitt Associates, cited in Jeff Goldsmith, "Death of a Paradigm: The Challenge of Competition," *Health Affairs,* 3(3:Fall 1984):p.12. Second two rows: National Center for Health Statistics, *National Health Survey,* various issues.

plan benefits, and some companies even cited the results in brochures to their employees explaining the changes. Thus it is plausible that the experiment played an important role in bringing about these changes.

The experiment, all told, cost around $80 million. Was the information worth that kind of money? Although $80 million is certainly a substantial sum, it pales by comparison with the $230 billion that the United States spent on hospital and physician services in 1984 alone. Table 3 shows that hospital discharges fell about 10% and physician visits fell about 4% between 1982 and 1984. If the results of the experiment caused even a small portion of this change—and few or no effects on health—then the experiment would have paid for itself in the space of a few months!

PROBLEMS

1. The following debate has occurred over the pricing of telephone service: Some people think that with a flat rate per month for local service, people spend much more time on the telephone than they would if they paid per minute. Others say the effect isn't very great. Do you think an experiment would be a good way to assess how people's use of the telephone might change if they had to pay by the minute? Why or why not? (Consider that some households now pay by the minute while other households pay a flat rate.)
2. Suppose you were going to conduct an experiment to assess how much telephone use responded to ending flat rate charges. Name some design decisions you would have to make. How would you induce people to participate?
3. Do you think an experiment would be the best way to assess the effect of the spread of health maintenance organizations on medical care prices and costs? Why or why not?
4. How does precision change with sample size? Is there a sample size that makes a study scientifically valid? Explain your answer.
5. When should the same number of people be assigned to each experimental treatment?
6. How should the number of sites in an experiment be chosen?
7. How long should an experiment run? Explain your answer.
8. What are the threats to the validity of experimental results from refusals to participate in an experiment?

REFERENCES

R. Archibald and J. P. Newhouse. In press. "Social Experimentation: Some Why's and How's." In E. Quade and H. Miser, eds. *Handbook of Systems Analysis, Vol. II.* New York: North Holland. (Also available as Rand publication R-2479, May 1980. Santa Monica, CA: Rand Corporation.)

R. H. Brook, J. E. Ware, Jr., W. H. Rogers, E. B. Keeler, A. R. Davies, C. A. Donald, G. A. Goldberg, K. N. Lohr, P. C. Masthay, and J. P. Newhouse. 1983. "Does Free Care Improve Adults' Health? Results from a Randomized Controlled Trial." *New England Journal of Medicine* 309:1,426–1,434.

W. G. Manning, J. P. Newhouse, N. Duan, E. B. Keeler, A. Leibowitz, and M. S. Marquis. 1987. "Health Insurance and the Demand for Medical Care: Results from a Randomized Experiment." *American Economic Review* 77:251–277.

Cigarette Price, Smoking, and Excise Tax Policy

Kenneth E. Warner *University of Michigan*

In 1986, Congress converted a temporary doubling of the federal cigarette excise tax into a permanent tax change. The principal motivation was the need for additional revenue. A second consideration was the belief that the higher tax would discourage hundreds of thousands of Americans from starting or continuing to smoke. In particular, Congress perceived that children and teenagers would be most influenced by the tax increase. If true, ultimately this could translate into avoiding many tens of thousands of smoking-related premature deaths and a great deal of smoking-related illness and disability.

Congress' expectations on both revenue yield and the consumption effects of the tax resulted from economists' studies of the price elasticity of demand, a measure of consumers' response to price changes. Examining data on cigarette consumption and price, over time and across states, economists have used a statistical technique known as multiple regression analysis (described below) to estimate the effect that price changes have on sales of cigarettes. These estimates, in turn, have been used by other analysts to predict the revenue and consumption effects of a tax change. The estimates of consumption impacts were then combined with epidemiological data on the health effects of smoking to estimate the number of premature deaths that would be avoided as the result of a tax increase, or caused by a tax decrease.

There is nothing unique about the use of regression analysis to estimate the elasticity of demand for cigarettes, nor of the use of the resultant estimates to influence policy. In the private sector, economists toil daily to relate consumer demand to product pricing. The price charged for a new car, for example, may be determined in part by regression analysis of whether a price increase (or decrease) will have a greater effect on unit revenue or on sales volume. Total revenue (*TR,* the total value of car sales) equals the price per car (that is, the unit revenue, P) multiplied by the number, or quantity (Q), of cars sold. Thus, for example, if a 2% decrease in price increases car sales by more than 2%, say, by 3%, total revenue will grow. This is because $0.98P \times 1.03Q > PQ$ because $0.98 \times 1.03 = 1.0094 > 1$, where $0.98P$ represents a 2% decrease in price and $1.03Q$ measures a 3% increase in quantity sold. If, in percentage terms, the price decrease exceeds the increased sales volume, total revenue will fall. In the previous example, if the 2% price decrease ($0.98P$) produces only a 1% increase in car sales ($1.01Q$), $0.98P \times 1.01Q < PQ$. (That is, $0.98 \times 1.01 = 0.9898 < 1$.) Whether the price or volume impact will be larger must be estimated by using the best data and statistical techniques available.

In this essay, we will examine how elasticity of demand is defined and calculated, how statistical analysis permits its estimation, and how it can be used to predict the tax revenue and cigarette consumption effects of a change in the federal cigarette excise tax. Drawing on published elasticity studies, we will produce specific estimates of the numbers of children, young adults, and older adults likely to have been discouraged from smoking by the 1986 decision to make the doubling of the federal tax permanent. And we will see how regression analysis and elasticity estimation can be used—and indeed were used—to inform both fiscal policy and health policy.

DEFINITION AND CALCULATION OF PRICE ELASTICITY OF DEMAND

The price elasticity of demand for a product is a measure of how much consumers change their demand for the product as a result of a change in the product's price. Technically, price elasticity is defined as the percentage change in the quantity of the product demanded in response to a 1% change in the price of the product.

While *elasticity* is a technical term in economics, its qualitative meaning is actually quite similar to what we mean by everyday use of the word *elastic*: A rubber band is elastic in the sense that it responds a lot (stretches) when pulled in opposite directions. In contrast, a pencil is inelastic in that it does not expand when both ends are pulled. Similarly, the demand for a product is elastic if it changes relatively more than the price that causes it to change (the price here being the exerted "force"). Demand is inelastic if it does not change as much in percentage terms as the price of the product. In the extreme, demand is perfectly inelastic if it does not change at all when the price of the product changes. This is analogous to the inelasticity of a heavy metal pipe when a person attempts to bend it using only bare hands.

Typically, demand for a good increases when the price of the good falls; demand decreases when the price increases. This relationship makes good intuitive sense: we all know that we want to buy more of something when it is less expensive; when it gets more expensive, we are inclined to buy less of it. Obviously we are more responsive to some prices than to others. For example, a 25% increase in the price of salt might not alter the amount of salt that we buy at all—our demand would be quite inelastic with respect to price; but a 25% increase in the price of a new car might cause us to postpone a new car purchase or to consider buying a used car or a less expensive new car than the one we had been considering. In this case, our demand for the car in question is quite price responsive, or elastic.

Calculation of the elasticity of demand is a simple matter if one knows the *demand curve* for the product. The demand curve is a graph showing the quantity of the product that consumers would buy at each of several different prices. Elasticity between any two points on the curve is calculated by measuring the percentage difference between the two quantities associated with the two points and dividing it by the percentage difference between the two prices associated with the two points.

ESTIMATING ELASTICITY OF DEMAND

The real world is rarely so kind as to present us with a well-defined demand curve. Rather, economists wishing to estimate the price elasticity of demand for cigarettes must find data on quantities of cigarettes smoked, prices, and other relevant variables (identified below) and then use them to estimate demand, in a form of statistical analysis known as multiple regression analysis (discussed below).

There are two basic approaches to estimating the demand for cigarettes. One, known as *time series analysis,* uses annual data on the variables of interest. In cigarette demand studies, national data have been used in time series analyses. The dependent variable in such studies is typically annual adult per capita cigarette consumption, defined as total cigarette sales divided by the size of the population over age 17. Cigarette sales data are available from the U.S. Department of Agriculture and are based on a variety of objective measures, including federal cigarette excise tax collections. Population data are available from the Bureau of the Census. The average retail price of cigarettes has been calculated by both governmental and nongovernmental agencies. A publication distributed by the Tobacco Institute (1988) provides detailed tax and price data and is used by many economists studying cigarette demand. Other annual data, such as income levels, are available from governmental sources.

The second basic approach to estimating demand for cigarettes relies on cross-sectional data. In a cross-sectional study, the analyst takes data from a single year for a number of different observational units. The most common unit for cigarette demand studies (as well as for many other demand analyses) is the state. Thus the analyst uses data on states' cigarette consumption levels and average retail prices (and average incomes, and so forth).

As all states impose their own excise taxes on cigarettes, each state has reliable data on tax-paid sales. While these data are an accurate measure of tax-paid sales, they do not necessarily accurately portray the actual levels of cigarette consumption by the states' residents because sizable differences in prices among the states, owing primarily to differences in excise tax levels, encourage cigarette smuggling —purchasing cigarettes in one (low-tax) state for use or resale in another (high-tax) state. Some smuggling represents the efforts of organized crime; but much of it is informal or casual bootlegging in which residents of one state living close to a low-tax state buy their own cigarettes in the neighboring state. Perhaps the most striking evidence of this phenomenon is the difference in tax-paid sales in Massachusetts and neighboring New Hampshire. In 1985, Massachusetts (with a cigarette tax of 26 cents per pack) had annual tax-paid sales of 117.2 packs per capita, while New Hampshire (17 cents per pack) had annual tax-paid sales of 201.1 packs, fully two-thirds greater than the national average (Tobacco Institute, 1988). There is no evidence that residents of New Hampshire smoke significantly more than residents of other states. There is evidence, however, of substantial interstate smuggling of cigarettes across the Massachusetts–New Hampshire border (Advisory Commission on Intergovernmental Relations, 1985).

Given the distortion that smuggling can introduce into cross-sectional studies, most analyses have relied on survey data from such sources as the National Health Interview Survey, conducted annually by the National Center for Health Statistics. These large surveys provide data on the health and health habits and demographic characteristics of Americans. By employing respondents' self-reported smoking habits, analysts believe that they are capturing actual smoking behavior better than if they relied on state tax-paid sales data. Analysts use the self-reported data on smoking behavior for the dependent variable, and use the state-specific price data as one of the independent variables (discussed below). To account for the possibility of cross-border bootlegging, recent sophisticated analyses also include neighboring states' prices when survey respondents live close to other states having considerably lower taxes. (See for example, Lewit and Coate, 1982.)

Surveys, however, are an imperfect source of data for a number of reasons. Foremost among these is people's tendency to underreport the amount of their smoking. One study found that, in 1975, respondents to a federal government survey reported less than two-thirds of all the cigarettes that were actually sold in the United States that year (Warner, 1978). Most cigarette demand analysts recognize this problem and address it in their studies.

As noted above, data such as these are used to perform a multiple regression analysis to study cigarette demand. A *multiple regression analysis* examines how a number of factors (called independent variables) relate to still another factor (the dependent variable). All of the factors take on a variety of values (hence the name *variables*). The assumption is that the independent variables vary "on their own" (thus the label *independent*), while the dependent variable is affected by (that is, "depends on") the movement of the independent variables. Thus a multiple regression analysis of the demand for cigarettes uses a measure of the quantity of cigarettes smoked as the dependent variable, and the quantity is said to *depend on* the values of the independent variables, which typically include the price of cigarettes and people's income levels. Other independent influences, often included as independent variables, are indices of antismoking publicity (which is believed to discourage smoking) and measures of recent previous smoking levels (reflecting the habitual or addictive nature of smoking). In each case, fluctuation in the independent variable is assumed to affect the number of cigarettes smoked or the number of smokers—the dependent variable.

Technically, multiple regression analysis, as with all statistics, identifies *correlations* between variables. As such, it cannot tell us that movement in one variable *causes* movement in another, but rather that the pattern of change in one is *related to* the pattern of change in the other. The function of multiple regression analysis is to sort out the multiple correlations among the independent variables and the dependent variable. To the extent that we can infer causality—or its likelihood—the regression analysis suggests how variation in each of the independent variables influences the value of the dependent variable. When, for example, we are interested in the relationship between quantity smoked and cigarette price, the analysis controls for (or fixes) the effects of the other independent variables and produces an estimate of the correlation between quantity smoked and price.

The results of a multiple regression analysis inform the analyst whether or not each independent variable is statistically significantly correlated with the dependent variable and, if so, how large the estimated effect is. Statistical significance means that the identified correlation is very unlikely to have occurred by chance. Technically, if the selected level of significance is .05, this means that the indicated correlation would occur by chance (that is, randomly) only 1 time in 20 if the variables are not related.

A statistically significant coefficient associated with the price variable estimates the amount by which the quantity of cigarettes smoked would decrease, given a one unit increase in price. By combining this estimate with the mean value of price and quantity, analysts can estimate the price elasticity of demand for cigarettes.

PREDICTING THE REVENUE AND CONSUMPTION EFFECTS OF A TAX CHANGE

Once we have an estimate of the elasticity and data on current consumption and price, it is relatively simple to predict the tax revenue and cigarette consumption effects of a change in the federal cigarette excise tax. For expository purposes, we will describe the effects of a tax decrease. The effects of a tax increase would be calculated in a precisely analogous manner, subject to replacing the word *decrease* with *increase,* and vice versa, throughout the discussion.

To estimate the consumption impact, recall that elasticity equals the consumption increase divided by the price decrease, both expressed in percentage terms. We know all of the essential information except the new level of consumption resulting from a tax decrease. The elasticity was calculated from the regression analysis. The percentage decrease in price is simply the amount of the tax decrease divided by the average price. Since we know the initial consumption level, we simply solve the elasticity equation for its one unknown, the new (post-tax) consumption level.

Estimating the tax revenue impact is also straightforward. The change in tax revenue accompanying a federal tax decrease is calculated by subtracting revenue in the year preceding the tax decrease from the estimate of revenue in the year following the decrease. In any given year, total cigarette tax revenue equals the tax per pack times the number of packs sold. In the year preceding the tax decrease, revenue equaled the original tax rate times the original quantity. In the year after the decrease, total revenue would be estimated as the new tax rate times the expected quantity estimated as just described.

Interest in the revenue impact of a tax change lies in the aggregate effect. Regarding consumption effects, however, interest may focus on subgroups within the population. That is, legislators (or health professionals or others) may want to know the impact of the tax on smoking by teenagers, since teens constitute the new generation of smokers and potential smokers. Alternatively, consumption effects of a tax change on older smokers may be of special concern since these are the individuals in whom smoking-related diseases will emerge soonest and hence they might derive the most rapid health benefit from quitting. Drawing on consumption data from surveys, economists can estimate demand functions for specific subgroups of the population, with age and sex the most common distinguishing variables. As we will see in the next section, this procedure has permitted economists to differentiate price elasticities for the various age groups and thereby makes possible estimation of differential consumption effects.

CONSUMPTION AND REVENUE EFFECTS OF THE 1986 FEDERAL CIGARETTE TAX MEASURE

The 1982 Tax Equalization and Fiscal Responsibility Act (TEFRA) included a doubling of the federal cigarette excise tax from 8 to 16 cents per pack, effective January 1, 1983. Defined as a temporary revenue raising measure, the tax was scheduled to remain in effect only until October 1, 1985, at which time the tax would return to the pre-TEFRA level. In part due to concern about losing the additional revenue, and in part out of fear that reducing the tax would encourage smoking, congressional legislators debated, and in early 1986 passed, legislation making the tax change permanent. (Between October 1 and the passage of the final legislation, the 16-cent tax was continued twice on a temporary basis.)

The legislators' understanding of the implications of the tax change benefited substantially from economic analysis that relied on studies of the demand elasticity of cigarettes to estimate the consumption and revenue impacts of permitting the tax's sunset provision to take effect. The analysis began with data on the national average retail price of a pack of cigarettes in 1984 (97.8 cents), a variety of measures of the quantity of smoking (including numbers of smokers in each of several age brackets and their annual consumption), and elasticity estimates for the various age groups, differentiating "participation" and "total" elasticities. Participation elasticities measured how decisions whether or not to smoke were influenced by price, while total elasticities assessed the traditional concern: how total cigarette consumption (the product of numbers of smokers and their average annual consumption) responded to price changes.

Table 1 presents estimates of cigarette demand and price elasticities derived by Dr. Eugene Lewit and his colleagues at the National Bureau of Economic Research. The overall adult elasticity (for ages 20–74), 0.42, means that a 10% increase in price would decrease consumption by adults by 4.2%. As can be seen in the table, and as seems eminently logical, older, more habituated (and likely more affluent) smokers are less price sensitive than are younger adults; that is, the elasticities fall as age increases. Consistent with this pattern and with the expectation, the overall elasticity for teenagers, 1.40, is higher than that of the youngest adults. For teens, a 10% price increase should decrease consumption by fully 14%.

From a health perspective, the participation elasticities are of greater interest, for they provide the "bottom line" indication of how many people would be influenced not to smoke by a price increase. The 0.26 participation estimate for all adults indicates that a 10% increase in price would discourage smoking by 2.6% of all adults who would have smoked in the absence of the price increase. Once again, participation elasticity falls as age increases.

Table 1 Cigarette demand and price elasticities

	ELASTICITIES	
Age Group (Years)	*Total*	*Participation*
12–17	1.40	1.20
20–25	0.89	0.74
26–35	0.47	0.44
36–74	0.45	0.15
All adults (20–74)	0.42*	0.26*

*Elasticities for all adults are calculated from separate regressions. This explains why the total elasticity is lower than the elasticity for each subgroup.

**Author note: The above tables are based on two separate studies, listed in source. These studied age group 12–17 in one study, and age group 20–74 in the other. Table 3 is based on an extrapolation between the two.

Source: Lewit and Coate (1982) and Lewit, Coate, and Grossman (1981).

Table 2 Expected percentage increases in cigarette consumption resulting from an 8-cent decrease in the Federal Cigarette Excise Tax

Age Group (Years)	*Total*	*Participation*
12–17	11.9	10.2
20–25	7.6	6.3
26–35	4.0	3.7
36–74	3.8	1.3
All adults (20–74)	3.6	2.2

For teenagers, the basic smoking decision seems to constitute the vast majority of the total elasticity. At 1.20, the teen participation elasticity means that a 10% price increase would discourage from smoking 12% of teens who would otherwise be smokers. Note that two groups comprise this 12%: teens induced to quit as the result of the price increase and teens discouraged from starting to smoke who otherwise would have done so. With the teen participation elasticity so high, price increases would appear to be a powerful tool to discourage children from adopting, or maintaining, tobacco habits.

The next step in converting the elasticities in Table 1 into estimates of the impacts of the specific federal tax change was a simple one: Analysts converted the originally legislated tax decrease into a price change. From a base price of 97.8 cents, the tax decrease of 8 cents would produce a new price of 89.8 cents, other things held constant (for example, assuming that manufacturers would not change the wholesale price). The percentage price change would be $8 / [(1/2) \times (97.8 + 89.8)] = 0.0853$, or 8.53%. (The denominator of the fraction is the average of the old and new prices.) If this figure is the denominator in an elasticity equation, the various elasticities from Table 1 can be inserted into the equation and the equation can be solved for the numerator, the percentage change in quantity demanded.

The results of these calculations are presented in Table 2. The table indicates that the 8-cent tax decrease would increase total adult cigarette consumption by 3.6%, and the number of adult smokers by 2.2%, while teenage smoking would increase by almost 12% (11.9%), including an increase in the number of smoking teens by 10.2%. Conversely, recision of the mandated tax decrease would avoid these percentage increases in smoking.

To convert the percentage change figures in Table 2 into numbers of smokers and cigarettes, we merely multiply the former by the numbers of smokers in each age category and their total annual cigarette consumption. Aggregate consumption data have been drawn from federal government surveys.

The results of this simple arithmetic are presented in Table 3. [The addition of two age groups (18–19 and 75+) in Table 3 and the underlying survey data are explained in Warner, 1986.] This table provides the answer to one of the two questions of interest to Congress: how many people would be encouraged to smoke if the legislated tax decrease takes effect; or, how many people would *not* smoke if the decrease were rescinded?

Table 3 Estimated increases in cigarette smoking attributable to an 8-cent decrease in the Federal Cigarette Excise Tax

Age Group (Years)	*No. Smokers (Thousands)*	*No. Cigarettes (Billions)*
12–17	334	2.3
18–19	130	1.0
20–25	609	4.8
26–35	508	4.1
36–74	351	8.2
75 +	13	0.2
Total	1,945	20.6

The numbers are impressive. Altogether, almost 2 million additional Americans would smoke if the tax decrease took effect. Of these, almost half a million (464,000) would be teenagers (age groups 12–17 and 18–19). More than half a million (609,000) would be in the youngest adult category, 20–25. In the category including middle-aged and older adults, the largest group of smokers, the numbers are smaller—351,000—but remain a substantial number of people. The analysis indicates that annual cigarette consumption would rise by 20.6 billion cigarettes.

All of these numbers appear to be quite large, including hundreds of thousands of people and tens of billions of cigarettes. In an absolute sense, the numbers *are* large. They reflect a lot of potential additional smoking and, with it, a considerable additional amount of smoking-related illness. But in *relative* terms, the indicated impacts are modest. Recognizing that some 56 million Americans are smokers, the participation effect of the 8-cent tax change amounts to only 3.5% of the smoking population. Similarly, the projected impact on total cigarette consumption is only about 5% of the cigarettes that people report that they smoke. Thus the consumption impact of the 8-cent tax change must be judged as modest, relative to the overall magnitude of the nation's smoking habit. But it is substantial in absolute terms: precisely because there are so many smokers consuming so many cigarettes each year, even a small percentage effect on smoking can translate into an important public health impact.

A crude estimate of that public health impact can be derived from the fact that one-quarter to one-third of lifelong cigarette smokers die as the direct result of their smoking habits (Mattson, Pollack, and Cullen, 1987). As such, if the 8-cent tax decrease took effect (and its real value were maintained over time), at least an additional half a million Americans would die prematurely as a result, although most of these deaths would occur several decades from now. Each would entail a loss of life expectancy of from 15 to 20 years. In addition to this most tragic outcome of a tax decrease, a substantial increase in smoking-related illness and disability could be anticipated.

Given the analysis of the consumption effects of a tax decrease, estimating the revenue implications of the mandated tax decrease is a trivial matter. Prior to the tax change, the tax per pack was 16 cents, and aggregate consumption was about 28.5 billion packs of cigarettes. Thus the total federal excise tax revenue, the product of these two figures, was $4.56 billion. If the tax fell by 8 cents, the new rate would equal 8 cents, while aggregate consumption would grow. According to Table 3, consumption would increase by 20.6 billion cigarettes, or just over a billion packs. But this estimate was derived from surveys, on which people have been shown to underreport their daily consumption by about a third, as discussed above. A better way to estimate the new level of cigarette consumption is to use the elasticity formula with an estimate of the overall (national) price elasticity of demand for cigarettes (produced by Lewit and colleagues) of 0.47. Doing this, we determine that the new consumption level would be 29.66 billion packs of cigarettes, an increase of 1.16 billion packs. Total excise tax revenue following a halving of the tax rate would be $2.37 billion. The tax loss accompanying a halving of the excise tax would be $4.56 billion minus $2.37 billion, or $2.19 billion.

As with the consumption impact, the revenue impact would be modest in relative terms (that is, relative to the $1 trillion federal budget) but more substantial in absolute terms. As the late Senator Everett Dirksen once suggested, if you take a billion dollars here and a billion dollars there, pretty soon it begins to mount up to real money.

Statistical elasticity estimation is nothing more than just that—estimation. The validity of the estimates *derived from* the elasticity studies is therefore dependent on the validity of the elasticity calculations. While it is impossible to confirm that validity with certainty, it is reassuring to note that in 1983, following the federal tax increase of 8 cents and manufacturers' price increases—a total price rise of 13.45% (after inflation)—total cigarette consumption fell by 5.13%. This implies an overall elasticity of 0.38, close to that estimated by Lewit and his colleagues. Their estimates look even better when we recognize that the smoking-age population grew from 1982 to 1983. Furthermore, the long-run impact projected by elasticity studies might well require more than one year to take full effect.

THE IMPACT OF ANALYSIS ON CONGRESSIONAL DECISION MAKING

No one would conclude that the above analysis played a major role in influencing Congress to rescind the legislated cigarette tax decrease. Political considerations—the classic "horse trading" for which Congress is justifiably famous—undoubtedly played a more significant part, as likely did the legislators' predispositions toward revenue matters in general and excise taxes (and cigarette taxes) more specifically.

Still, anecdotal evidence suggests that this analysis did sway the thinking of certain key actors in the congressional debate. One senator reportedly had planned to filibuster during the debate on the bill to rescind the tax decrease. He felt that Congress had made a commitment in 1982 to return the tax to its pre-TEFRA level and that it should uphold its commitments. According to an aide to the senator, however, the senator was so moved by the realization that the tax decrease would cause almost half a million children to smoke that he dropped his plans to filibuster.

Public health lobbyists who followed the debate closely reported that several other members of Congress were also influenced by this finding, which was published in a leading medical journal, presented at a Senate hearing, and covered by the media. Thus, while it is impossible to assess the precise contribution of the economic analysis to the enactment of the permanent 16-cent tax, it is reasonable to conclude that economic analysis, and its underlying statistical analysis, served to inform the congressional decision-making process and to inject a note of statistically based objectivity into a proceeding that was otherwise highly politicized.

PROBLEMS

1. Explain what a "price elasticity of demand" measures. In what way(s) can elasticities assist in decision making?
2. What is a "multiple regression analysis"? Describe how it is used to estimate price elasticities.
3. The overall price elasticity of demand for cigarettes is inelastic. Should profit-maximizing cigarette manufacturers want to raise or lower their product prices? Explain.
4. Assume that the overall price elasticity of demand for cigarettes is 1.50, instead of its actual estimated value of 0.47.
   a. Estimate the aggregate consumption and tax revenue effects of permitting the excise tax to decrease from 16 to 8 cents per pack.
   b. Under the actual situation of inelastic demand, both revenue and public health considerations recommended maintaining the tax at 16 cents. Given your calculations in (a), would health policy and fiscal policy work together if the elasticity were 1.50? Explain your answer.
5. The implicit assumption in the statistical estimation of elasticities is that responses to price increases and decreases are symmetrical. This assumption —used in this essay for expository simplicity—leads to the conclusion that a 10% *increase* in cigarette price would result in a 4.7% *decrease* in the demand for cigarettes, just as a 10% price *decrease* would produce a 4.7% demand *increase.* Many economists believe, however, that a cigarette price increase would have a proportionately larger effect on demand than would a price decrease.
   a. Explain why this asymmetry seems logical in the case of smoking.
   b. Would you expect the asymmetry to be comparable for teenagers and middle-aged adults? How does your answer relate to the differential demand elasticities in Table 1?

6. Elasticity estimates derive from regression analyses. Different analyses—employing different data or functional specifications—can and do produce different elasticity estimates. Would an opponent of the 1986 recision prefer to use a higher or lower elasticity estimate to argue against maintaining the tax at 16 cents? Discuss.

REFERENCES

Advisory Commission on Intergovernmental Relations. 1985. *Cigarette Tax Evasion: A Second Look.* Washington, D.C.: Advisory Commission on Intergovernmental Relations.

E. M. Lewit and D. Coate. 1982. "The Potential for Using Excise Taxes to Reduce Smoking." *Journal of Health Economics* 1:121–145.

E. M. Lewit, D. Coate, and M. Grossman. 1981. "The Effects of Government Regulation on Teenage Smoking." *Journal of Law and Economics* 24:545–569.

M. E. Mattson, E. S. Pollack, and J. W. Cullen. 1987. "What Are the Odds that Smoking Will Kill You?" *American Journal of Public Health* 77:425–431.

Tobacco Institute. 1988. *The Tax Burden on Tobacco—Historical Compilation.* Washington, D.C.: Tobacco Institute.

K. E. Warner. 1978. "Possible Increases in the Underreporting of Cigarette Consumption." *Journal of the American Statistical Association* 73:314–318.

K. E. Warner. 1986. "Smoking and Health Implications of a Change in the Federal Cigarette Excise Tax." *Journal of the American Medical Association* 255:1,028–1,032.

Does Inheritance Matter in Disease? The Use of Twin Studies in Medical Research

D. D. Reid* *London School of Hygiene and Tropical Medicine*

The controversy about the relative importance of nature and nurture that goes on in fields such as psychology also goes on in medicine. The crucial question is: Do we go mad or develop heart disease because we inherit a special susceptibility to a disease from our parents, or are these diseases the result of a stressful environment throughout our lives? With most diseases, we cannot decide because patients almost always differ in both their nature, or genetic endowment, and their nurture, or environment, during the formative years of childhood, and of course, both can be important in particular diseases. Yet it is important to know whether in coronary heart disease, for example, it is worth persuading patients to alter their environment by stopping smoking. If they both smoke heavily and have heart disease because they have inherited an anxious, worrying temperament, then altering their smoking habits alone might have

*Deceased.

disappointingly little effect. For these reasons, medical research in this area has concentrated on the unique and unusual opportunities presented by the occurrence of disease in twins.

The idea for this approach came from the English biometrician Francis Galton in 1875. He pointed out that there are two kinds of twins, identical and nonidentical. Identical twins come from the same fertilized egg and therefore have the same set of genes, which determine their physical and mental characteristics. Nonidentical twins, like ordinary siblings born at different times, come from separate eggs and their gene patterns are no more (and no less) alike than those of ordinary brothers and sisters (see Figure 1). Both types of twins, however, usually share the same family environment during their childhood and are usually treated alike by their parents. If, then, one of the pair develops a disease that is transmitted through the genes, it is more likely to appear in a twin who is identical than in a twin who is genetically different. On the other hand, in diseases due to diet or some other aspect of family life that is not genetically determined in the strict sense, identical and nonidentical twin siblings of affected children are equally likely to develop the disease. Thus we can get some indication of the relative importance of heredity and environment in causing a specific disease by comparing the relative risk of a twin being affected when he or she is an identical twin to that risk when the twins are nonidentical. The difference in risk will be high in diseases where inheritance is important and low where it is not.

MEASURES OF DISEASE "CONCORDANCE"

Risks of the coincidence of disease appearing in both twins are measured in two ways. Both methods aim to assess the degree of *concordance,* or agreement, between the disease experience of identical and nonidentical twins. Table 1 shows a model population of six sets of identical twins. One measure of risk of both being affected is the "pair-wise concordance rate." In this

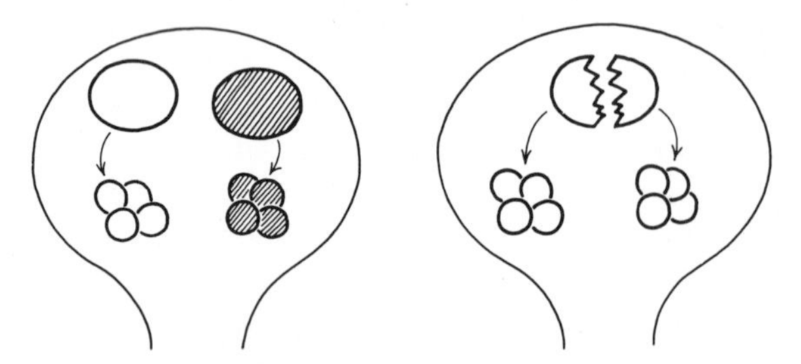

Figure 1 *Genes determining light hair are indicated by ○s and dark hair by ●s. Nonidentical twins come from separate eggs that carry different genetic elements that determine, for example, hair color in the two children. The color may or may not be different in the case of nonidentical twins. Identical twins result from the division of the same egg and so retain the same genetic element. Both children thus have the same hair color.*

Table 1 Concordance example

Pair Number	*First Twin*	*Second Twin*
1	⊙	⊙
2	⊙	⊙
3	⊙	⊙
4	⊙	⊙
5	⊙	○
6	⊙	○

(○ = unaffected; ⊙ = affected.)

example, in four out of the six pairs, both twins are affected, so the pair-wise concordance rate is $^4/_6 \times 100\%$ or $66\frac{2}{3}\%$. When neither twin has the disease the pair does not enter the table. The second measure depends on the relative frequency or risk of the twin of an affected person also suffering from the disease. This "proband concordance rate," or "affected concordance rate," in the example is given by the ratio of the number of twins affected as pairs (8) to the total number affected (10), that is, $^8/_{10}$ or 80%. The affected concordance rate is perhaps the more widely used rate in twin studies. (Knowing one of these measures, we can readily obtain the other, so no important choice is being made here; it is more a question of custom.)

SOME RESULTS FROM THE DANISH TWIN REGISTRY

Table 2 compares the affected concordance rates for identical and nonidentical pairs of Danish twins of whom at least one is affected by one of the diseases listed. The high rates in identical twins versus the lower ones for nonidentical twins for tuberculosis (54 versus 27), rheumatoid arthritis (50 versus 5), bronchial asthma (63 versus 38), and epilepsy (54 versus 24) suggest a strong genetic element in these diseases. On the other hand, death from acute infections other than tuberculosis and rheumatic fever (14 versus 11) shows no such disparity between the identical and nonidentical pairs of twins. Chance exposure to

Table 2 Occurrence of selected somatic diseases in the Danish Twin Register based on a survey of 4,368 same-sexed pairs

	Affected Concordance Rates (%)			
Disease	*Identical Rate*	*No.**	*Nonidentical Rate*	*No.**
Cerebral apoplexy	36	120	19	164
Coronary occlusion	33	122	27	179
Tuberculosis	54	185	27	309
Rheumatic fever	33	178	10	238
Rheumatoid arthritis	50	63	5	73
Death from acute infection	14	137	11	235
Bronchial asthma	63	94	38	125
Epilepsy	54	37	24	49

*Numbers refer to number of affected individuals, not pairs.

infection, perhaps outside the home, seems therefore, to be much more important than any inherited susceptibility.

THE UNUSUALNESS OF TWINS

These examples have shown the potential of this method of distinguishing between genetic and environmental factors in different diseases. A word of caution is needed, however, about the risks of generalizing from twins to the population as a whole, for twins are unusual in more senses than one. They are unusual in that multiple births occur relatively infrequently. In 1964 only 1 delivery in 96 was of twins. (Triplets and quadruplets occurred once in 9,977 and once in 663,470 deliveries—compared with the ratios of 1 in 9,216 and 1 in 884,730 expected on the basis of the Helin-Zeleny hypothesis; that is, if twins occur once in 96 deliveries, triplets and quadruplets will occur once in 96×96 deliveries and $96 \times 96 \times 96$ deliveries, respectively. This hypothesis does not concern us here.)

Most important from the medical point of view is the fact that twins are more likely to be born to black than to white Americans and especially to older mothers who have already had several children. In a condition like mongolism, which is particularly common in children born to older women who have had large families, twins are thus likely to be more often affected than are singletons. Because twins have to depend on nutrients designed for a single child in the womb, they start at a disadvantage and are more likely to be born prematurely and to have difficult births. It is hardly surprising that their death rate in infancy and early childhood is higher than average. As they grow older, their disadvantage lessens, and in respect to the diseases of adult life, their experience is probably close to that of singletons. Thus assessments based on observing disease in middle-aged twins are likely to be reasonably applicable to people in general.

PRACTICAL ASPECTS OF TWIN STUDIES

Sampling Populations of Twins. Early studies of disease in twins were often based on a patient with an unusual disease who came to the hospital and was found to have a twin suffering from the same condtion. As in clinical medicine in general, the apparently unusual tends to be noted and published. Series of such coincidences in twins have thus been given prominence in the medical literature. But, because of the haphazard method of collection, such series are unlikely to give a true picture of the incidence of disease in the population of twins as a whole. Volunteer series, recruited perhaps by appeals in the media, are also likely to be biased. The ideal is to collect information either on all twins born in a generation or at least on a randomly selected, and thus truly representative, sample of them. Particularly in Scandinavia where vital records have long been accurate and complete, national twin registers have been established on this basis to serve medical and social research.

Some results from the Danish Twin Register have already been given. This Register comprised all twins born in Denmark during a defined period (1870–1910). Of the 37,914 twin births that occurred during that time, over half of the pairs had been broken by the death of one or both twins before their sixth birthday; these were not followed up. About 40% of the pairs were of different sex and not so useful for investigating diseases occurring in one sex more than another. Of the remainder, some 60% were nonidentical twins of the same sex and 40% were identical and so like-sexed pairs. (In other words, the elimination of mixed-sex pairs from the surviving sets of twins changed the identical-nonidentical ratio of twin births from the usual 20:80 to 40:60.)

Establishing Type of Twins. Having identified and traced the twins through the population registers and other means, a problem arises in establishing their type by methods that can be widely and simply applied. In studies of small numbers, refined techniques can be used to compare inherited characteristics such as blood groups or blood-protein patterns to see whether in all such respects the twins are truly identical. Fingerprints, voice sounds, and other physical traits can also be used. For large-scale surveys, in which subjects cannot be examined but only interrogated by postal questionnaires, simpler methods are needed. It is fortunate, therefore, that the reply to one question is surprisingly effective in distinguishing identical from nonidentical pairs of twins. That question may take the form of "Were you as like as two peas in a pod?" and it is encouraging to note that when this question was asked, over 95% of pairs in which both answered yes proved, on blood and other examination, to be identical, and when both answered no, over 95% were fraternal; finally, 2% disagreed.

Once the pairs of twins have been classified, their disease experience must be ascertained. This can be done by collecting hospital records or death certificates over a period of years or by asking questions directly about either past illnesses or the presence of symptoms of chronic diseases such as coronary heart disease or rheumatism.

Other Applications of Twin Studies. Twin studies also may help to detect the effects of environmental factors in disease. Cigarette smoking, for example, is believed to cause other lung diseases as well as lung cancer. As in the case of lung cancer, it could be argued that an individual inherits both a liability to take up smoking and a specific susceptibility to lung diseases such as chronic bronchitis. If this were true, the apparent association between cigarette smoking and the presence of bronchitis could be dismissed as an effect arising at least in part from other causes rather than as proof that smoking caused bronchitis.

Surveys of smoking habits in identical and nonidentical twins have shown that there is indeed some evidence of a genetic element that affects the smoking habit. Regarding smoking as a disease, we can, as before, compare the affected-concordance rate in identical twins with that in nonidentical twins. This shows that, if one twin smokes, his or her twin is more likely also to be a smoker if they are identical twins than if they are fraternal (and thus no more closely related than an ordinary brother or sister).

Table 3 Prevalence of cough among smokers and nonsmokers in smoking discordant twin pairs

	Cough Prevalence (%)		*Total Number of Cases*
	Smokers	*Nonsmokers*	
Identical twins			
Men	14.6	7.7	274
Women	13.6	7.6	264
Nonidentical twins			
Men	12.3	5.5	733
Women	14.5	5.7	653

Because of this genetic element in smoking, the independent effect of smoking on bronchitis has to be assessed by comparing the frequency of bronchitis in identical twins who share the same genetic endowment but smoke different amounts. A survey based on the Swedish National Twin Study has done just this and the results for both types of twins are given in Table 3. These are set out simply in the form of prevalence rates that give the percentage of people in each group who have chronic coughs. Clearly, within each group of identical twins, smokers have higher prevalence rates for chronic cough than nonsmokers. In other words, even when the genetic background is identical, smoking appears to be associated with more bronchitis and thus is likely to be its cause.

Related Ideas. The use of twins is not confined to studies of medicine or biology; they have been used in studies of reading, for example. The idea of using identical twins as a control in the smoking investigation is a special case of the important statistical idea of using homogeneous "blocks" to test different effects. First, keeping close to the twin idea, in agricultural experiments, littermates are sometimes assigned to different treatments such as feeding regimes to improve the precision of the resulting average weight gains under the different diets. Of course, many litters may need to be used, but fewer individuals will be needed because of the "matching" provided by the litters.

Future International Collaboration. The results from the Scandinavian Twin Registries show how useful such data can be. Unfortunately, for less common diseases even large national registries may not uncover enough cases of a specific disease (as in different forms of heart disorder) to make detailed statistical analysis possible. The World Health Organization has therefore set up a "registry of registries" to collect data in a uniform fashion in many countries and assemble and analyze them centrally. In this way we hope to reap the full benefit of the unique research opportunities that studies of twins in sickness and health can provide.

PROBLEMS

1. Why are volunteer series in twin studies likely to be biased? (*Hint:* Refer to the essay by Meier.)

2. By comparing the risk of an identical twin being affected to that of a nonidentical twin, an experimenter would eliminate any ________________ factors, and thus any observed differences between the two kinds of twins could be attributed to ________________ factors.

3. In the data of Table 3, only those pairs of twins are considered for which:
   a. Either both are smokers or nonsmokers.
   b. One twin is a smoker and the other a nonsmoker.
   c. Either both or one or none are smokers.

   Which answer is correct?

4. Refer to Table 3. Can you convert the data of cough prevalence from percentage into incidence (number of cases)? Carry out this conversion for the identical twins if you can; explain what other information you need if you can't.

5. Explain how one should use twins in studies to detect:
   a. The effect of genetic factors in disease.
   b. The effect of environmental factors in disease.

REFERENCES

M. G. Bulmer. 1970. *The Biology of Twinning in Man.* New York: Oxford University Press.

A fuller account of "The Use of Twins in Epidemiological Studies" is given in a WHO report in *Acta Genetica et Gemellologia* 15(2, 1966): 109–128. The Danish Registry and its work are described in the same journal, 18(2, 1968): 315–330. The report "Multiple Births USA 1964" from the National Center for Health Statistics is published by the U.S. Public Health Service.

The Plight of the Whales

D. G. Chapman *University of Washington*

Between the end of World War I and 1960, several species of whales in the ocean around the Antarctic continent were the basis of an important industry. These giant mammals, the largest that have ever existed on the earth, were sought for animal oil and, to a lesser extent, meal and meat (the latter for human consumption) as well as a myriad of by-products. In antiquity, whalers went out in small boats and endured great risks to capture such large sources of meat. Men continued to hunt whales in small boats with primitive weapons, as portrayed in *Moby Dick,* until late in the nineteenth century. In the twentieth century, whaling has been highly modernized with explosive harpoons, large ships, and powerful radar-equipped catcher boats, which enable the whaling industry to operate in the stormy and inhospitable oceans next to the Antarctic ice cap.

This area of the world, while unfriendly to humans, is very inviting to whales, for during the southern summer the waters bloom with small plants that feed myriads of minute animals, known generally as krill. Certain species of whales catch these by straining large volumes of water in their huge mouths through sievelike filters called baleen plates (hence this group of whales is referred to as baleen whales). These whales have no teeth and do not eat fish or other marine mammals. The largest of the baleen whales, and indeed of all whales, are the blue whales, which may reach a length of 100 feet, though 70 to 80 feet is a more usual size.

BLUE WHALES

Immediately following World War II, Europe and Japan were in desperate need of many things, including animal oil. It was not surprising, therefore, that the number of Antarctic whaling catcher boats increased; furthermore, technologies developed during the War made whaling more efficient. As a result, some conservationists feared that Antarctic whales, particularly the blue whales, would be completely eliminated. Figure 1 shows the annual catch of blue whales in the southern oceans in the decade before the War and in the postwar period to 1960. The basis for concern for the blue whales was easy to document, but the catch of other species was stable or increasing. Some of those associated with the industry suggested reasons other than a decline in population for the decline of the blue whale catch and were reluctant to accept restrictions on catches. Thus the International Whaling Commission, set up in 1946 to manage the resource, found itself the center of controversy. The Commission has representatives from all interested countries; it establishes size regulations and quotas, and it also has the authority to ban hunting of species that appear to be endangered.

The Commission set up a study group to bring together all the data and develop statistical methods for attacking such questions as: How many whales are in the stock that feed in the Antarctic? How many young are born each year?

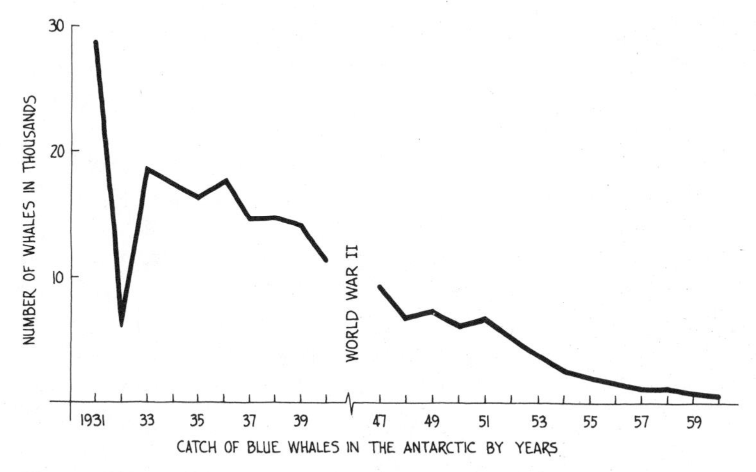

Figure 1 *Catch of blue whales in the Antarctic, seasons 1930–31 to 1959–60 excluding World War II period.*

How many whales die from natural causes each year? How are these birth and death rates affected by factors over which humans have some control?

COUNTING METHODS

Let us consider the first, and perhaps most basic, question: How many whales are there of a particular species? Whales unfortunately don't stay still to be counted. They roam over large areas, spending most of their time under water, though they do surface at regular intervals to breathe. Furthermore, the southern oceans cover a vast part of the world; the whaling area exceeds 10 million square miles, an area larger than all of North America. There are several standard ways to estimate wild animal populations, all of which involve some statistical techniques. We shall describe three of them.

Marking Method or Capture–Recapture. The first method of estimation involves marking a number of whales; a foot-long metal cylinder is fired into the thick blubber that lies just under the skin. If and when marked whales are later caught, some information is available on their movement, on their rate of capture, and on the proportion of marked members in the whole herd. The usefulness of the latter information is easily seen by simulating such an experiment with a can of marbles. Assume that, like the whale population, the number of marbles in the can is unknown. Now pick a few marbles (say, 10) out of the can, mark them, and return them to the can. Next, stir the whole can thoroughly and draw another sample. Count separately the marked and unmarked marbles. If the unmarked ones are four times as numerous in the sample as the marked ones, we reason that the same is true of the whole canful; but because there is a total of 10 marked marbles, we infer there are 40 unmarked marbles, or 50 marbles in total.

This simple scheme has been used with many animal populations, although there are many obvious complications in practice, and for whales this is especially true. How do we know, for example, that the metal mark fired into the blubber actually penetrated and did not ricochet off? Did the crew who cut up the captured whale carefully look for the mark—even a foot-long metal cylinder is easy to overlook in cold, stormy working conditions when the volume being cut up is approximately the size of a house. Also, unlike the marbles, whales are born and die over a period of years. All of these complications require refinements and extensions of the simple experiment outlined here. It is necessary to have a series of experiments extending over many years and to use comparative procedures. For example, if a group of whales is marked in year 1 and a group of the same size is marked in year 2, then, after year 2, the ratio of recoveries of whales marked in year 1 to the recoveries of whales marked in year 2 reflects the proportion of marked whales of group 1 that died in the intervening year. These deaths may have been natural or caused by hunters. Moreover, the *ratio* is a valid measure of this mortality because its numerator and denominator are equally affected by the possible errors listed above. Such a comparative study is only one of the several statistical procedures used to analyze whale-marking data.

Catch-per-Day Method. The second estimation method is based on changes in the rate of catching whales. The rate of catching depends mainly on the frequency with which whales are seen and, other things being equal, this depends on their density. Thus the catch per day reflects the density. How can this be translated into absolute numbers? If the change in catch per day is entirely a result of the removal by man, then it is easy to make this translation; if catching 25,000 whales in one season lowers the catch rate for the next season by 10%, then at the outset of the first season there must have been 25,000/0.10, or 250,000 whales.

Again, the situation is more complex than this simple example. Whaling ships hunt over a vast area in difficult conditions, so that the catches fluctuate violently. Whaling companies introduce new technology to improve their efficiency. Moreover, we reemphasize that there are other causes of whale mortality, and that there are new births as well; both of these factors must be taken into account in adjusting the population estimate. One way to overcome some of these difficulties is to adjust for changes in efficiency and also to follow the change in catch per day (adjusted) over a period of several seasons. Figure 2 shows the catch per day of blue whales plotted against the cumulative catch by the whaling factory ships over the seasons 1953–1954 to 1962–1963 when natural deaths and births were numerically quite small. As more whales were caught, the catch per day went steadily down. This graph suggests that there were only 10,000 to 12,000 blue whales in 1953 and this number declined to about 1,000 in 1963. As pointed out, there are statistical refinements, and the result obtained in this way must be combined with estimates obtained in other ways.

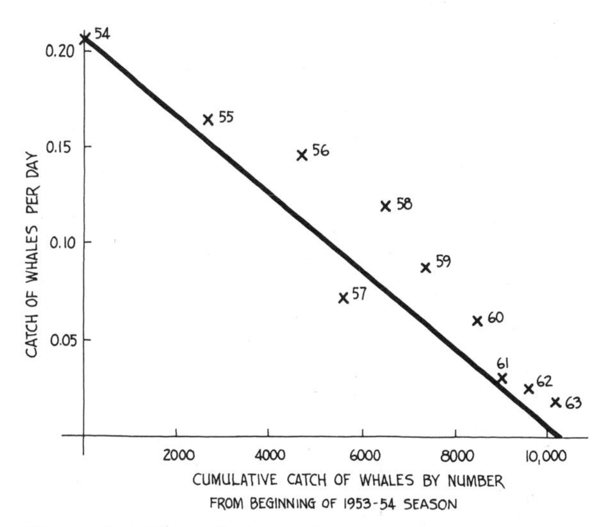

Figure 2 *Blue whale catch per day (adjusted for efficiency improvements) versus cumulative catch, 1953–54 to 1962–63.*

Age Analysis. The catch-per-day method works well with rapidly declining populations, but in other situations the complications and corrections make it less useful. Still a third method is available, however, which uses the ages of whales. Just as trees have annual rings in their trunks and fish have annual rings in their scales, whales have annual rings in a waxy secretion in the ear (earplugs). The ages of a sample of the whales killed each year were determined by the rings of their earplugs. In addition, information on the length of every whale killed commercially made it possible to relate age to length and to calculate an estimated age for every captured whale.

It was thus possible to make a statistical estimation of the number of 4-year-old whales in any season and the number of 5-year-olds in the following season. Because one year's 5-year-olds are the survivors of the previous year's 4-year-olds, a survival rate or, conversely, a mortality rate can be determined. Because all ages are estimated and because some adjustments have to be made, the estimated mortality rates fluctuate wildly. But by averaging over several year's classes, over areas and seasons, useful results can be obtained. Furthermore, with additional statistical analysis, it is even possible to assess the magnitudes of possible errors in such estimations. These mortality rates help us to predict the future of the whale population.

Results. Thus we have three methods of estimating population sizes and mortality rates: the marking method, the catch-per-day method, and age analysis. The results of different methods were checked against one another, and fortunately the different estimates were in good agreement. Sources of error were carefully checked and ruled out, so that the study group finally concluded that the blue whales numbered at most a few thousand and might total even less than 1,000. Thus there was and is a real danger of extinction of this species in the Antarctic (there are also small numbers of blue whales in the northern oceans). Fortunately, the International Whaling Commission banned the taking of blue whales as soon as the study was finished—first, in a large part of the southern oceans and eventually in all waters south of the equator. It is too soon to predict the long-term survival of the species; blue whales are occasionally seen, but these are probably the survivors noted above (whales can live, in the absence of hunting, to more than 40 years of age). We can ask whether the population has been reduced to such low levels that reproduction is reduced below the level necessary for species continuation, but it will be a number of years before this can be answered.

FIN WHALES

The second-largest whale species in the world, also part of the baleen family, is the fin whale. It averages about 10 feet less than the blue whale in length. During the 1950s this stock annually yielded in excess of 1 million barrels of oil per year. With the decline of the blue whales, fin whales bore the brunt of the exploitation. The same methods of analysis used for the blue whales were applied to the fin whales; in fact, the analysis was more critically needed because the condition of the fin whale stock was not obvious as was that of the blue whales. Moreover, fin whale catches were still very high: in the 1961–1962 season over 27,000 fin whales were killed. The study group recommended that the fin whale catch should be reduced to 7,000, or less, if the fin whale stock was not to be further depleted. The proposal for such a drastic reduction came as a shock to the Commission; the study group forecast that the next season's catch, regardless of quotas, would drop to 14,000. When actual figures proved the forecast right, most countries wanted to move toward the drastic reductions required, but some of the whaling nations were able to block action. Another disastrous season caused a revision in the thinking of the commissioners, and in 1965 a substantial schedule of reductions in the quota was agreed upon. Nevertheless, the delay in reaching this agreement, and the subsequent delay in reducing the quota, meant that permitted catches have had to be lowered even further, and, in fact by the mid-1970s the Commission gave total protection to fin whales in the Antarctic and in the North Pacific.

MORATORIUM ON COMMERCIAL WHALING

By the time the major species, blue and fin whales, were placed in protected status, Antarctic pelagic whaling was carried on by only two countries, Japan and the U.S.S.R. These countries turned to other species, first the sei whale and then the minke whale. The latter is, by comparison, a very small whale. Its maximum length is about 30 feet. Little was known about these species—only a few had been marked. The methods of determining their age are uncertain and controversial. Furthermore, scientists had no long series of data on catches so that the data base for determining the status of these stocks was totally inadequate. Thus through the late 1970s and early 1980s the International Whaling Commission found itself in great controversy. Many felt that it should not permit whaling of stocks on which so little was known. Also, new countries that favored a conservative approach joined the Commission. Finally, in 1982, the Commission voted to establish a moratorium on all commercial whaling not only in the Antarctic but worldwide. In passing this resolution the Commission also agreed to undertake by 1990 a comprehensive assessment of all whale stocks to hopefully resolve the uncertainties that had given rise to the controversy.

THE FUTURE OF WHALE STOCKS

Some stocks of whales have now been protected for nearly 25 years, yet there is no information available to determine whether they are beginning to recover. If the comprehensive assessment is to resolve the status of such seriously depleted stocks and the more recently exploited ones, new methods of obtaining and analyzing information need to be found. Most of the information now available on whales was collected from dead animals taken in commercial kills. Such data has its limitations and in any case is not available during the moratorium. New methods include sighting expeditions from ships or aircraft. While considerable work has been done to extrapolate such sighting information to determine population numbers, other biological information is more difficult to collect in this way. Other potential methods are photography and skin samples. Whales have natural marks that can be identified visually or from photographs and these can be used as a basis of mark recapture procedures. This can substitute for the marks used heretofore (the metal cylinders), which are recovered only when the whale is killed. All of this opens up new areas of research both in techniques and in statistical analysis of the data so acquired.

In the meantime, the whole question of killing whales for commercial purposes is being seriously debated. Some agree that whales are a resource that should be harvested like fish or trees, and further, only by harvesting whales can humans efficiently harvest the great krill resources of the oceans. Others assert that whales are intelligent animals with a complex social structure and that whaling is not only cruel but also destroys this social structure. Such persons believe that preserving whales is symbolic of our intent to live in harmony with the natural world.

PROBLEMS

1. Refer to Figure 1. What was the approximate catch of blue whales in the Antarctic in the season 1931–1932? 1937–1938? What is the percentage reduction between the two seasons? When was the catch the highest? The lowest?
2. What are the three counting methods discussed in the text? Describe each briefly.
3. Describe an experiment whose purpose is to estimate the size of a human population on an isolated island, using one of the methods mentioned in the article. What assumptions are you making?
4. Refer to Figure 2. What was the approximate catch of blue whales per day (adjusted) in the season 1955–1956 (plotted as 1956)?
5. Refer to Figure 2. Can you explain the relative "high" for the 1958 figure?
6. From Figure 2 it was concluded that there were only 10,000 to 12,000 blue whales in 1953. How was this number obtained?
7. Suppose it was determined by the age analysis method that the mortality rate of blue whales is 0.25 per year. Furthermore, suppose that by the marking method it was estimated that there were about 7,000 blue whales at the start of 1960. How many whales would you expect to be still living at the end of 1961, assuming that the mortality rate does not change over the years?
8. What is meant by the "optimum stock size" of whales? Why should there be an optimum size at all?

REFERENCES

D. G. Chapman, K. R. Allen, and S. J. Holt. 1964. "Reports of the Committee of Three Scientists on the Special Antarctic Investigations of the Antarctic Whale Stocks." *Fourteenth Report of the International Whaling Commission.* London: International Whaling Commission, pp. 32–106.

J. A. Gulland. 1966. "The Effect of Regulation on Antarctic Whale Catches." *Journal du Conseil,* 30:308–315.

N. A. Mackintosh. 1965. *The Stocks of Whales.* London: Fishing News.

Scott McVay. 1966. "The Last of the Great Whales." *Scientific American* 215(2): 13–21.

The Importance of Being Human

W. W. Howells *Harvard University*

In the summer of 1965, my paleontological colleague, Bryan Patterson, was in charge of a Harvard expedition working near the shore of Lake Turkana in northern Kenya. At a locality called Kanapoi, while he was searching in deposits believed to be of the early Pleistocene (the last geological epoch before the present), he picked up an important fossil. The broken lower end of a left humerus (the upper arm), it was easily recognized as ***hominoid***; that is, it came from a creature of the group that includes humans and their closest living relatives, the apes, but not from a monkey.

What was the special importance of the fossil? From shape and size it could be seen at once not to belong to a gorilla, an orangutan, or a gibbon (and the last two have never been present in Africa anyhow). It was extraordinarily similar to the same piece in modern humans; in fact, it was indistinguishable. But the date of the deposit was certainly before the existence of anything like modern humans, and after the field season was over, volcanic basalt from a bed lying above the deposit gave an age estimate, by radioisotope dating, of about 2½ million years. The oldest human stage that had been established so far was that of the erect-walking but small-brained and large-jawed australopithecines found by Louis Leakey at Olduvai Gorge, which had been dated at about 1¾ million years. If this small piece of arm bone were "human," or ***hominid,*** in the sense of belonging to such a creature, it would extend the continuous record of human evolution backward three-quarters of a million years at a single bound.

But there was one problem. This piece of elbow joint in humans can easily be told from that in orangs, gorillas, and gibbons, but not from that in chimpanzees. Although the rest of a chimpanzee's bone is shorter and stouter, this region is so similar in the two species that many, if not most, specimens defy classification as one or the other on examination. In spite of different uses of the arm, this particular part shows such slight, subtle, and inconstant distinctions in size and shape as to baffle ordinary methods of study even by experts. The problem, therefore, was this: either the bone was that of the earliest australopithecine yet found in our direct ancestral history or it was simply that of an ancestral chimpanzee, in which case we could breathe normally. What about testing something old with something new? Could an electronic computer tell us anything useful?

A computer, of course, does not really "tell" anything. It merely makes possible answers to mathematical questions that we would not live long enough to answer if we tried to work them out with simple calculating machines. With its enormous capacities and speed, a computer transfers the effort from getting the right answer to getting the right question. Biological material—bones or skulls are good examples—lends itself to particular kinds of questions. Because the genes they inherit are capable of a virtually infinite number of different combinations, no two individuals of a population or species are exactly alike (with the spectacular exception of identical twins). So, quite apart from different habits of use, diet, or other accidents of growth, human elbow joints vary normally in size and details of shape, though they vary within a limit of form that is basic to the actions of human elbow joints.

Quite different species of animals, of course, have quite different forms in various body parts. Any beginner can distinguish between a cheek tooth of a mammalian carnivore, with its narrow, knifelike shearing crown, and that of a herbivore, which has a broad surface for grinding vegetable matter. These are marked evolutionary divergences. Within herbivores the differences are smaller, and within groups of herbivores such as pigs (for example, domestic pigs, wild boars, warthogs, etc.) or elephants, species distinctions are matters for experts, who can obtain a wealth of information from fossils about the history of pigs and elephants or about the exact species of animals present at a given time in the past at a fossil locality such as Kanapoi. Finally, for particular parts, such as the elbow joint in chimpanzees and humans, the species distinctions may be so slight as to be eclipsed by the variation ***within*** each species, already described. That is the situation we are faced with here.

This is not just a matter of impression: it may be viewed quantitatively. Some time ago, Professor William L. Straus of Johns Hopkins University, a man with much experience in such studies, tried to deal with the same problem when the same piece of the humerus of a species of australopithecine was found at the site of Kromdraai, near Pretoria, South Africa. In this case, it was plain that the bone belonged to *Paranthropus,* the species in question, because other skeletal and cranial parts of the same species had been found at the site, and the bone could hardly be assigned to anything else. Here the problem was whether the bone was more humanlike or more apelike, since the hominid ("human") position of the australopithecines at that time was less clear. Professor Straus made a number of typical measurements of human and chimpanzee bones in an attempt to find differences between them. He found statistically significant differences[1] in the averages of certain of the measurements, but the absolute differences were slight, and the overlap in each measurement between human and chimp was so great that the *Paranthropus* fragment could not be allocated to either. In no case did its measurements lie outside the range of either human or chimpanzee, though the figures were more often closer to the mean, or average, figures for the latter.

This was no solution and led to no decision as to the relationships of *Paranthropus,* insofar as the arm could shed light on them. In such a case, we need a method that is not limited to comparisons of single measurements, that somehow takes account of the whole shape of the bone, or part, as the eye tries to do, and that also has some way of emphasizing the really telling differences in shape between two species, if they exist. Now here is an important point: in the end, any such problem comes down to a mathematical question because the eye itself (though very seldom consciously) attempts to assess the *average* differences in proportions and complex aspects of shape, to rate the varying importance of these, and, finally, to judge the probability that a given total shape, in a single case, falls nearer to the essential basic form within the variation of one population than to that of another. These are questions of quantity and probability, whether measured or not, and are thus statistical in nature. After all, educated opinion is always the weighing of probabilities. And here is another important point: biologists and anthropologists—and members of many other sciences—are not often strong in mathematics of a higher order, though they may see only too acutely the limits of their own ways of solving problems. At the same time, mathematicians, although they have hearts of gold, are not usually sufficiently conversant with the niceties of biological problems to understand just what the biologist is trying to gain by using a mathematical analysis. When the two really get together, however, the rewards in the way of new solutions may be great. And I must say that mathematical training among biologists who see better what such training can offer has increased notably in recent years.

Fortunately, the particular problem of the Kanapoi fossil is not exceedingly complex, and the solution was provided some years ago by the great English

[1]The reader may recall that, when statisticians can detect a difference likely to be a real effect and not one stemming from chance variation, they call it "statistically significant." By saying "statistically" they warn us that the absolute size of the difference may be small and seemingly unimportant because it depends on the objects being studied.

statistician and geneticist R. A. Fisher in the form of the *discriminant function.* The discriminant function eliminates the futile business of looking at measurements one at a time, of finding that the overlap prevents discrimination of two sets of specimens, such as human and chimp elbow joints, even though they are known to be from quite different animals, and of being unable to place something such as the *Paranthropus* specimen logically nearer one group than the other. It has a set of weights with which to multiply a number of different measurements of a specimen, the sum of the products being a single *discriminant score* that makes the best attainable use of all the information in the several measurements. Given two groups, such as humans and chimpanzees, the computation develops the optimum set of weights possible from the measurements used: the effect is to sift out important differences—often quite invisible to the eye or in average figures—so as to emphasize precisely the aspects of shape and size that will best discriminate between the two groups. That is, compared to just that variation *within* a set of human elbow joints, or chimpanzee elbow joints, the distinctions *between* the sets are searched out mathematically so that the discriminant scores of the two groups are segregated one from the other to the maximum degree possible, limited only by the information contained in the measurements. Thus the overlap, acting as a mask to hide any real group differences, is reduced or removed.

The basic idea of the discriminant function may be appreciated graphically in the case of *two* measurements, represented by the two axes of Figure 1. (The measurements might be heights and girths of two different groups of men.) The oval areas correspond to groups of individuals from two populations A and B. If we look just at measurement 1 by dropping projections on the horizontal axis, we find considerable overlap between the two populations. The same holds for measurement 2. On the other hand, the slanted line perfectly separates the two populations. This is not the place for mathematical detail, but to write the previous sentence is to say that looking at something like

$$(\text{Measurement } 1) + 2 \times (\text{Measurement } 2)$$

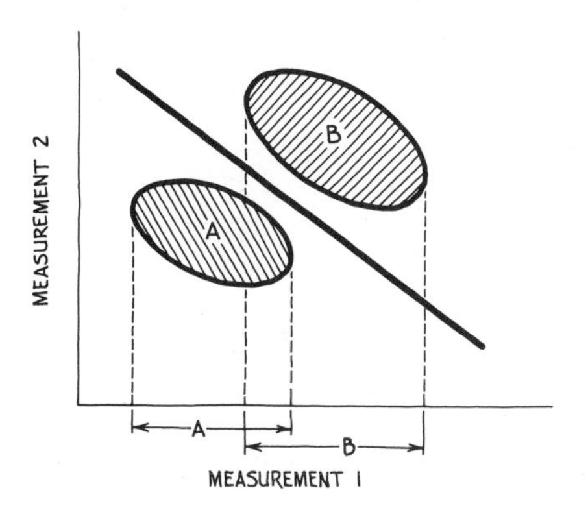

Figure 1 *Two measurements together separate groups better than either separately.*

gives us a new score, the discriminant score, which permits much better separation of the populations than either of measurements 1 or 2 alone. If there are more than two measurements, as in the present case, then there are great potential gains in combining measurements.

Professor Patterson and I felt fairly strongly that the Kanapoi fragment was hominid—on the human, not the ape, side of the hominoid group as a whole. But we wanted to demonstrate this statistically, not merely to voice an opinion to which opposing opinions could be raised by others. As a strategy, we examined human and chimpanzee humeri to see what measurements would most likely reflect such differences as we thought appeared, whether frequently or not. Figure 2 shows the fragment itself and some of the measurements. To begin with, we took the total breadth across the whole lower end, as a matter of general size (measurement 1). Second, the more projecting inner, or medial, epicondyle (at the left in the figure) has a snub-nosed, or slightly turned-up, aspect in some chimpanzees, and we hoped to register this effect by measuring from the lowest point on the trochlear ridge both to the "beak" of the epicondyle and to the nearest point on the shoulder just above it (measurements 2 and 3). The idea was that a slightly greater difference between these two would reflect a deeper curve and more upturned epicondyle. We also measured the backward protrusion of the central, or trochlear, ridge of the joint, the length and breadth of the oval inner face of the medial epicondyle (none of these is shown in Figure 2), and an oblique height of the opposite, or lateral, epicondyle. We thought these measurements showed some tendency to vary one way in humans, the other in chimpanzees, though not being the rule in either (if there *were* regular distinctions, obviously the problem of discrimination would be much less). We were not certain of the functional meaning of the possible differences, but they logically could be related mostly to muscle attachments connected with simpler and more powerful use of the flexor and extensor muscles of the hand in the chimp, in hanging by the arms or supporting the body in ground-walking by the characteristic resting on the middle knuckles, all as contrasted with the more general, but more complex and varied, use of the hands in humans.

Now this is just where the cooperation comes in. It is the paleontologist's or anthropologist's business, from his or her background knowledge, to find

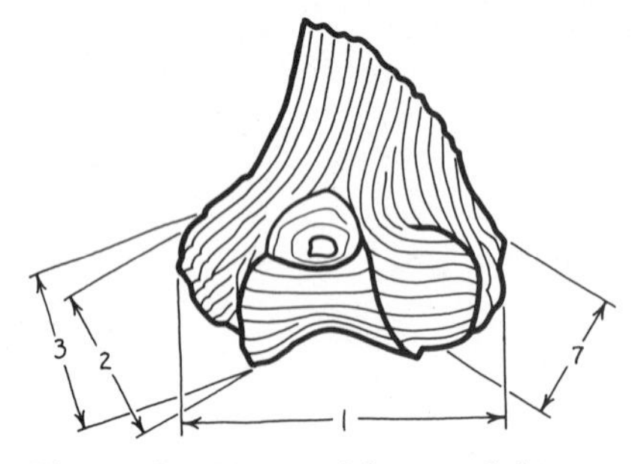

Figure 2 *Kanapoi humeral fragment and measurements taken.*

measurements that will carry important and real information as to differences. It is the statistician's business to say how the measurements can be put together to bring out the differences for evaluation. Here, cooperation had already gone so far in recent years that the biologists knew in advance what statisticians could offer them and we planned our work accordingly.

We measured 40 human bones in the Peabody Museum at Harvard and 40 chimpanzee bones in the Harvard Museum of Comparative Zoology and the American Museum of Natural History in New York. As in Straus's measurements, the overlap of human and chimp was great, but the mean differences, resulting from the special selection of measurements, were in most cases better defined. The means for the two groups and the figures for both the Kanapoi and Kromdraai fragments (the latter taken on two casts) are given in millimeters in Table 1.

The chimpanzee specimens, as a sample, may be accidentally a little large on the average. The *Paranthropus* fragment is obviously small in all dimensions and so appears "human" when we glance at this list; however, this does not necessarily mean that the shape relations conform to those of humans. The correspondences of the Kanapoi measurements to the human means (of this particular sample) are very close throughout—closer than we might expect any random human bone to be in all its measurements.

To assure ourselves of this apparent closeness, we computed a discriminant function from the human and chimp figures. For only seven measurements and such small samples, the calculations could be done by hand, though at the cost of no little labor. In technical language, matrices have to be formed of the sums of all the cross multiplications of all the measurements of all the individuals both within each group and of the total lot; other steps require the inversion of one matrix and the determination of the latent roots of another. Inversion by hand of a matrix of even the modest size of 7×7 is a tedious business and one open to error. This all leads to finding the discriminant function, which takes the seven measurements from a specimen, multiplies each measurement by a weight specific to that measurement, and then adds these products to give the discriminant score. This is a great deal of arithmetic, and we can only say that to have a computer handle such a job in a matter of minutes is very welcome. Waiting for paint to dry or for a film to be developed now seems long and drawn out by comparison, and such easy computation has obviously greatly encouraged undertakings such as the one described here.

Table 1 Measurements

Measurement	*Chimp Mean*	*Human Mean*	*Kanapoi*	*Paranthropus*	*Scaled Vector*
1. Bi-epicondylar width	64.1	58.0	60.2	53.6	− .09
2. Trochlea-medial epi. distance	44.8	40.7	41.7	33.6	+ .40
3. Trochlea-supracondyl dist.	41.3	38.8	39.4	32.1	− .62
4. Posterior trochlear edge	26.4	22.1	22.2	19.9	+ .11
5. Medial epi. length	24.7	20.3	20.8	15.5	+ .19
6. Medial epi. breadth	12.8	12.6	13.9	10.4	− .32
7. Lateral epi. height	31.5	26.7	27.6	24.9	+ .56

The last column in Table 1 gives not the actual weights in the discriminant function as used, but rather a rescaled form of the weights with their relative importance in proper perspective (because, for example, a small measurement, such as thumb length, might require a much larger weight in the function than a large measure, such as stature, to make it effective). These figures show how a number of measurements combine to form a single pattern of greatest difference between the two groups. As might have been expected, the two measurements to register the snub-nosed effect of the medial epicondyle, or its opposite, are useful, as shown by the large size of the scaled vector values. The plus value of measurement 2 and the minus value of measurement 3 combine to make the total discriminant score higher when the epicondyle is most turned up; that is, when measurement 2 is high relative to measurement 3 (see Figure 1), the function creates a greater positive value to add and a smaller minus value to subtract in the total score, and when the opposite is true, with the shoulder of the condyle more sloping, there is on balance a greater minus value in the total score. The lateral epicondyle (measurement 7) also adds a greater plus value when it is high, while the breadth of the medial epicondyle (measurement 6) adds to a plus value (or rather subtracts *least* from a total value) when it is relatively narrow.

Table 1 shows that the above are indeed characteristic human-chimp differences in the averages (though small ones), all of which tend to produce higher score values for the chimpanzee. We note that there is almost no ***absolute*** difference in measurement 6, the breadth of the medial epicondyle, certainly not a significant one, and yet this measurement is important in discrimination because it is ***relatively*** narrow in chimpanzees, whose other measurements (in these samples) are larger, on the average, than the human measurements.

We notice also that because the discriminant score is affected by all measurements, it takes account of variation in form toward or away from a basic pattern: if a chimpanzee bone lacks any snubbing of the medial epicondyle, it may exhibit another combination of narrow epicondylar face or high lateral epicondyle, and so it may score in a chimpanzee direction anyhow.

When the discriminant scores were calculated, they produced a far greater separation of human and chimpanzee bones than did any of the measurements singly. Here are the mean score values, and the limits of the individuals in each group:

	Mean	*Range*
Chimpanzee	99.77	67–130
Human	61.42	40–84

All but two of the chimpanzee values fall between 80 and 120, and all but one of the human values fall between 50 and 75, which are nonoverlapping intervals. So the separation was very good: of 80 specimens, only 3 overlapped, falling closer to the wrong mean figure than to their own. Unquestionably, this is a successful procedure to distinguish human and chimpanzee humeri by measurement, with a much greater probability of correct assignment than is possible by eye.

Now for the scores of the Kanapoi and *Paranthropus* fragments. These were 59.4 and 63.9, respectively, very close to the human average (almost too good to be true, being closer than most of the known human individuals) and, of course, outside the range of the 40 chimpanzee values entirely. Using statistical theory we compute that, had either bone actually belonged to a chimpanzee, it would have a discriminant score as small as those above (or smaller) with a probability of only about 1 in 500. With so small a probability, we conclude that the two bones did not come from chimpanzees, but from hominids, and that is the answer to the question we framed.

Of course, we must be careful. The real question (because of the material we used) was this: how do the fragments classify themselves when they are asked to choose between *modern* human and *modern* chimpanzee arm bones? These were the only alternatives which we offered to fossil creatures which existed when there were no modern humans, and when ancestral chimpanzees might also have differed significantly from those of today. Nevertheless, we have good grounds for inferring from their shape that the arm bones were used, on the whole, like those of humans and, at least, not like those of the African apes, terrestrial though they are to a great extent. This takes care of the Kanapoi individual and, as a bonus, says the same thing for the hitherto baffling fragment from Kromdraai.

To review: unable to establish from visual inspection that the Kanapoi fossil did not belong to an animal like a chimpanzee, we turned to measurement and a statistical procedure that could be applied with the help of a computer. (As biologists, we knew from experience how to state the problem and extract a great deal of information, but how to order, analyze, and judge the information we learned from statisticians.) Though moderately complex, the discriminant function is well suited to the biological realities of individual variation and group differences in shape and gives an answer that states a numerical probability from the known evidence. So, by the middle of 1966, Professor Patterson and I had concluded that we could rule out the possibility that he had found the bone of an ape and that from what we know about East Africa, the only other possible possessor of the fossil was an early hominid, that is, an australopithecine.

Much more is now known in 1988 about australopithecines and about East African paleontology. The Kanapoi formation is dated at more than 4 million years ago, not 2½ million. On a later expedition, Professor Patterson found a piece of a lower jaw, at Lothagam Hill west of Lake Turkana, dated at about 5 million years ago. This has been diagnosed as very probably hominid, as have a couple of other minor fragments found elsewhere in the same region. There is now a general consensus that hominid origins lie between 5 and perhaps 8 million years in the past.

Also, a considerable amount of skeletal material has been recovered elsewhere in East Africa that is more than 3 million years old, especially in the Ethiopian sample made famous by "Lucy." But the Kanapoi fossil is still the earliest such limb fragment, and it is still decidedly modern in form—as confirmed by later and more complex analyses—compared to Lucy and her companions. In any case our original conclusion, that it represents an australopithecine, is sustained and now appears much less surprising. And we may take note that our 1967 statistical analysis was decidely helpful in affirming the importance of the Kanapoi fossil, before the later material was in hand.

PROBLEMS

1. In the hypothetical example of Figure 1, project the values of measurement 2 for groups A and B onto the vertical axis and indicate the interval where the two populations overlap on measurement 2. Use Figure 1 to explain why we need to look at both measurements (or a function of these measurements) to be able to classify an individual as being in A or B.

2. Suppose instead of the two populations in Figure 1 we have the populations in Figure 3 below. It is then very easy to discriminate between the two populations. How would you do it?

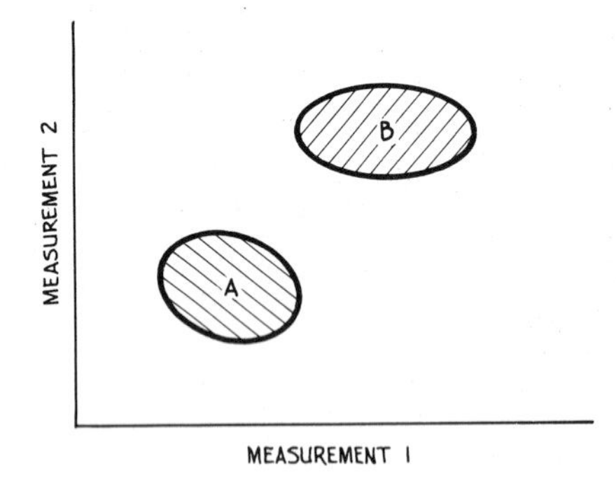

Figure 3

Use this example and the one in Figure 1 to explain when discriminant functions are needed.

3. In Table 1 note that there is almost no difference in measurement 6 for humans and chimpanzees, yet it was one of the measurements that helped to discriminate between the two populations. Explain how measurement 6 might help in disciminating between the two populations.

4. Give a detailed example of your own in which a discriminant function analysis might be appropriate.

PART TWO
Our Political World

Statistical Proof of Employment Discrimination

Sandy L. Zabell *Northwestern University*

Title VII of the Civil Rights Act of 1964 prohibits discrimination in employment on the basis of race, color, religion, sex, or national origin. In part because of the passage of the Civil Rights Act, and in part because of changes in the social climate, discrimination of the type prohibited under the Act is usually not as blatant as it once was, and proof of its existence is correspondingly much harder than before the Act's passage. During the last decade the courts have turned increasingly to the methods of statistics to assist them in assessing the existence and extent of discrimination.

The U.S. Supreme Court has played a key role in the judicial adoption of statistical methodologies to prove employment discrimination. Although statistical proof began to appear sporadically in the lower courts almost immediately after passage of the Civil Rights Act, during the past two decades the discussions of statistical proof in certain key Supreme Court decisions has played an important part in the willingness of the lower courts to accept such proof and in their increasing reliance on it.

DISPARATE TREATMENT AND DISPARATE IMPACT

Employment discrimination cases fall into two broad categories: *disparate treatment* and *disparate impact.* In a disparate treatment case, the issue is whether an employer has intentionally discriminated against someone. In such cases statistics can be introduced as circumstantial evidence to establish a "pattern and practice" of discrimination (as was done in *Hazelwood,* described below). In *Griggs* v. *Duke Power Company* (401 U.S. 424, 1971), however, the Supreme Court ruled that employment practices called "facially neutral," such as tests for hiring and promotion, were also discriminatory under Title VII if they had an adverse or disparate impact on a protected class. Even when an employer does not intend to discriminate, if the employment practices result in the hiring or promotion of protected classes of individuals at a substantially lower rate, a prima facie case has been made that the employer is in violation of the Act. By divorcing *outcome* from *intent,* the Supreme Court created an essentially statistical category of discrimination, and that in turn made the use of statistical proof inevitable.

Proving that a facially neutral employment practice has a disparate impact does not necessarily suffice to prove prohibited discrimination. Under the "business necessity" defense, an employer can defend a practice by showing that it has a legitimate business purpose. If, for example, it can be shown that an employment test is testing for a legitimate job skill, then this can serve as a defense. Validating an employment test (showing that it does test for such a skill) is also an area where statistics can be used, but, as a practical matter, it may often be hard for an employer to do such validation in an adversarial setting, and thus the demonstration of a substantial statistical disparity will often suffice to decide a case.

HAZELWOOD

Although the Supreme Court had on several earlier occasions pointed to statistics as a valuable tool in employment discrimination cases, its use of statistical tests of significance in *Hazelwood School District* v. *United States* (433 U.S. 299, 1977) gave the formal methods of statistical proof a stamp of approval they had not enjoyed previously. Since *Hazelwood* the lower courts have felt free to use statistical methodology in deciding employment discrimination cases.

In *Hazelwood,* the U.S. government brought suit against the Hazelwood School District, alleging that it had engaged in a "pattern or practice" of employment discrimination in violation of Title VII. Hazelwood is a largely rural school district in the northern part of St. Louis County, with few blacks in its student body and few (under 2%) black teachers on its staff. The government based its statistical case on the fact that the percentage of black teachers in the district was substantially smaller than in the surrounding area of St. Louis County and the city of St. Louis.

The district court (where the case was first tried) compared the percentage of black *teachers* in the district with that of black *students,* noted that they were roughly the same, and ruled in favor of the Hazelwood School District. When the case was reviewed, the Federal Court of Appeals ruled that the comparison of teachers with students was irrelevant and that the appropriate comparison was the percentage of black teachers in the *school district* with that of the black teachers in the relevant *labor market* (just as the government had argued), and reversed the lower court decision. The case was then reviewed by the Supreme Court.

The Supreme Court's decision notes that of 405 teachers hired by the Hazelwood School District during the two school years 1972–1973 and 1973–1974, 15 (or 3.7%) were black, whereas the proportion of black teachers in St. Louis County was 5.7%. One of the questions facing the Court was whether these two percentages were substantially different.

One way to begin analyzing such a question is to calculate the probability that a disparity this large or larger could occur. If one regards the 405 hires as a random sample from a pool of qualified teachers containing 5.7% blacks, then the probability of selecting 15 or fewer blacks may be calculated to be 4.6%. Such a probability may be calculated directly by using a *binomial probability distribution* (see Mosteller, Rourke, and Thomas, 1970, Chapter 4).

Such direct calculations are often tedious, and in practice the significance of such a disparity is often judged by an approximate procedure that involves computing a basic statistical quantity called the *standard deviation.* This measures how variable the data are. One then notes how many standard deviations the observed number of individuals (here 15) is from the expected number (here $405 \times 0.057 = 23$). A basic rule of thumb is that if individuals *are* selected at random, then 95% of the time the observed number will lie within 2 standard deviations of the expected number (more than 2 standard deviations above occurs about 2.5% of the time and more than 2 standard deviations below about 2.5%). Conversely, if the discrepancy exceeds 2 standard deviations, then a comparatively rare event has been observed and this calls into question the assumption that individuals were chosen at random; the difference is then said to be statistically significant at the 5% level.

The formula for the standard deviation in this setting is $\sqrt{np(1 - p)}$, where n is the number of individuals, and p is the proportion in the population. For example, in *Hazelwood,* $n = 405$ and $p = 0.057$, so that the standard deviation in this case is $\sqrt{(405)(0.057)(0.943)}$ or 4.67. In a footnote (number 17), the Supreme Court noted that the difference between the number of blacks hired and the number expected ($23 - 15 = 8$) was less than 2 standard deviations, which would indicate that the difference is not statistically significant.

An important question that arose in *Hazelwood* was identifying the appropriate comparison population or relevant labor market. If one includes the city of St. Louis along with St. Louis County (as the government had argued), the percentage of black teachers increases from 5.7% to 15.5%. (This was in part because the city of St. Louis school system had attempted to maintain an approximately 50% black staff.) Obviously in this case the disparity would achieve statistical significance. Because the disparity between 3.7% and the smaller labor market percentage of 5.7% was less than 2 standard deviations, the Court concluded that in this case the difference might be "sufficiently small to weaken the government's other proof, while the disparity between 3.7% and 15.5% may be sufficiently large to enforce it." Since this issue had never been considered by the lower courts, the appellate court decision was reversed, and the case was remanded to the district court for retrial.

Thus the value of statistical evidence as proof in an employment discrimination case can depend in a very sensitive way on which "relevant" labor market is chosen as a standard for comparison. In scientific sampling, where a random sample is drawn from an unambiguous and prespecified population, this standard of comparison must, of necessity, be clearly identified prior to drawing the sample. But in cases such as *Hazelwood,* what one begins with is the sample—the hiring rate for a protected class—and the population or relevant labor market is a hypothetical construct from which the hired employees are imagined to have been drawn. As the text of the *Hazelwood* decision makes clear, the choice of the relevant labor market to be used as a standard of comparison can be a complex question not necessarily admitting of a unique, simple, or unambiguous answer.

When the disparities observed in an employment discrimination case do not exceed those readily accounted for under random sampling, the result is some evidence against the presence of discrimination. If, on the other hand, a disparity is flagged as statistically significant, such a result needs to be interpreted with caution. For one thing, a hiring process need not be random in a statistical sense in order for it to be fair or unprejudiced. Fair processes of hiring can correspond to methods of sampling that will average to the correct expected value, but have larger standard deviations than those of simple random samples (see Meier, Sacks, and Zabell, 1986). For another, "statistical significance" merely means that there is *some* difference, but not that this difference necessarily has practical or legal significance. A difference of one or two percentage points in either direction, for example, in the pass rates of men and women on a physical dexterity test would scarcely be remarkable. As the appellate court in *United States* v. *Test* (550 F. 2d 577, 10th Cir. 1976) observed, "The mathematical conclusion that the disparity between these two figures is 'statistically significant' does not, however, require an *a priori* finding that these deviations are 'legally significant.' "

TWO-SAMPLE COMPARISONS OF APPLICANT DATA

Because deciding on the relevant labor market can involve delicate issues of judgment, the courts have expressed a strong preference for comparing applicant data with applicant data whenever possible (for example, what percentage of female applicants are hired versus the percentage of male applicants). This avoids having to decide on what the relevant comparison population is and addresses a much more relevant question: Given the actual applicant pool, how did the protected class do? In the previous section we compared the percentage in one group with a well-determined population value, and thus dealt with a one-sample problem. Here we are comparing percentages of female and male applicants hired and so we are comparing the outcomes for two samples and we must allow for the variability from both sources. We discuss this shortly.

For example, in *Connecticut* v. *Teal* (457 U.S. 440, 1982)—another Supreme Court case—a written examination used to determine eligibility for promotion was challenged on the ground that it had a disparate impact on black examinees. Of 48 blacks who took the test, 26 passed (54.2%), while of 259 whites who took the test, 206 passed (79.5%). In this case the question is whether, if in the long run blacks and whites pass the test at the same rate, a sample difference this large or larger could occur. In *Teal,* the difference in pass rates is 79.5% − 54.2% = 25.3%, which is 3.76 standard deviations, well in excess of the "2 or 3 standard deviations" benchmark enunciated by the Supreme Court in its *Hazelwood* decision.

In many cases involving small businesses or short periods of time, the samples involved are too small to justify the use of the approximation that lies behind the standard deviation benchmark. Then recourse may be had to the ***Fisher Exact Test*** (see Mosteller, Rourke, and Thomas, 1970). The Fisher Exact Test views those individuals hired or promoted as a random sample from the applicant pool and calculates the probability that a disparity in hire or promotion rates at least as large as the one observed could arise.

Statistical analysis in cases involving small numbers can be particularly helpful because on many occasions intuition can be highly misleading. It is often thought that with small samples no useful information can be extracted, but it all depends on the size of the disparity involved. For example, in ***Dendy*** v. ***Washington Hospital Center*** (431 F. Supp. 873, D.D.C. 1977), the following table resulted:

	Selected	*Rejected*	*Pass Rate*
Blacks	4	5	44%
Whites	26	0	100%

The difference in pass rates for the two groups is obviously substantial, but because there are only 9 blacks in the entire sample, intuition might suggest that the difference could easily have happened by chance. In fact, such a disparity is extremely unlikely: In ***Dendy,*** the significance probability resulting from the Fisher Exact Test is .0004 (that is, there are no more than 4 chances in 10,000 of a difference this large or larger arising when treatment of the two groups is alike). (The history of ***Dendy*** is interesting: At the district level, the court—Judge John Sirica of Watergate fame presiding—rejected the statistical evidence on the ground that the sample involved was too small, but he was later overruled when the case was reviewed at the appellate level.)

Although consideration of applicant data avoids the problem of determining the relevant labor market, it can suffer from its own problems. Paradoxically, affirmative action programs aimed at encouraging minority groups to apply for positions can lead to applicant pools with differing levels of ability. In *Washington* v. *Davis* (426 U.S. 229, 1976), for example—another Supreme Court case—a test used by the District of Columbia Police Department was ruled to be nondiscriminatory, despite clear evidence that it had a disparate impact, partly because the pools may have had differing levels of ability.

SIMPSON'S PARADOX AND AGGREGATION

Large corporations and companies are almost always divided into divisions, subdivisions, departments, and so on. Except for high-level managerial positions, employment decisions usually take place at the departmental or divisional level. Analyzing aggregate employment data in such companies can give rise to a curious phenomenon known as *Simpson's paradox* (sometimes also referred to as *spurious correlation*).

An instructive and surprising example of Simpson's paradox occurred at the University of California at Berkeley in the 1970s. Examination of applicant data for a 1973 quarter revealed that the overall rate of admission for female applicants to the graduate school was substantially less than the rate of admission for male applicants (see Table 1).

Which departments at Berkeley were responsible for this imbalance? Surprisingly, admission data for individual departments showed that admission rates for males and females were comparable; indeed, in some departments admission rates were significantly *higher* for women!

This paradoxical situation actually has a very simple explanation. At Berkeley, women had applied more often to departments, such as English and History, with large numbers of applicants and correspondingly low rates of admission, while men had applied more often to departments, such as Mathematics and Physics, with fewer numbers of applicants and much higher rates of admission. A simple hypothetical example will illustrate the phenomenon:

	Mathematics			*English*			*Combined*		
	Admit	*Deny*	*%*	*Admit*	*Deny*	*%*	*Admit*	*Deny*	*%*
Males	90	10	90	1	9	10	91	19	83
Females	9	1	90	10	90	10	19	91	17

In this example, the admission rates for men and women are the same in each department: 90% in Mathematics, 10% in English. Nevertheless, because women applied more often to English and men more often to Mathematics, and because the overall admission rates for the two departments differed, the aggregate admission rates for men and women were substantially different.

Table 1 Applicants for graduate admission to the University of California, Berkeley, Fall 1973

Sex of Applicant	*Number of Applicants*			*Percentage Admitted*
	Admitted	*Denied*	*Total*	
Male	3,738	4,704	8,442	44%
Female	1,494	2,827	4,321	35%

Source: Bickel, Hammel, and O'Connell (1975).

Because of Simpson's paradox, it is appropriate that employment data in discrimination cases be analyzed at the level where decision-making actually takes place. Despite its appealing simplicity, considering only aggregate data can be highly misleading: fair employment practices can be made to appear unfair, and unfair employment practices can be disguised.

DISCUSSION

The use of statistics in employment discrimination cases reveals a surprising spectrum of issues involving application and interpretation, only some of which we have been able to explore here: the relevant comparison population, the appropriateness of the test of statistical significance based on random sampling, effects of aggregation, and controlling for relevant explanatory variables such as skill, to list but a few. The failure to take such problems into account can often result in a highly misleading picture, as in Simpson's paradox. Nevertheless, carefully used and appropriately interpreted, statistical methods provide some useful tools for the courts in their task of determining when prohibited discrimination in hiring, promotion, or firing has taken place.

PROBLEMS

1. In a recent case involving a suburban police department, a physical performance examination was challenged on the ground that it discriminated against women. While approximately 80% of 150 male applicants passed the exam during a six-year period, only 3 of 10 women passed during the same period. If the disparity is real, how could use of the examination be defended?
2. In the Berkeley graduate admissions case, could the different admission rates for departments be used as an argument for the presence of discrimination? Argue pro and con.
3. In 1968 Dr. Benjamin Spock and several associates were tried in Boston for conspiracy to violate the Military Service Act of 1967 (see *United States* v. *Spock,* 416 F. 2d 165, 1st Circuit 1969). Although more than 50% of adults in the Boston area were female, there were no women on Dr. Spock's jury!

This reduction came about in three stages: due to statutory disqualifications only 30% of the eligible jurors were female (stage 1); the jury list of 100—from which the jury was selected—contained only 9 women (stage 2); and—as a result of peremptory challenges—the final jury contained no women at all (stage 3). This was a matter of considerable concern to Dr. Spock's attorneys since women were thought to be favorably inclined toward their client, both because of his well-known books on child care and because polls indicated that women were more opposed to the Vietnam War than were men.

If you were an attorney for Dr. Spock, how would you attack the progressive diminution of women at each stage in the jury selection process? In particular, for stage 2, how large is the discrepancy in standard deviation units when a list of 100 jurors, drawn at random from a pool containing 30% women, contains 9 or fewer women?

4. In both employment and jury discrimination cases, the usual statistical calculations presuppose that the individuals in question (the persons hired or the jurors selected) are a random sample from the applicant or potential juror pools. In what ways other than discrimination might this assumption be violated in each case? At what stages in jury selection does the assumption seem more reasonable in jury discrimination cases?

5. From 1974 to 1978 the federal income tax rate for individuals decreased in every income category, yet the average overall rate of taxes actually collected increased from 14.1% to 15.2% (Wagner, 1982)! Explain why this is—in disguised form—an instance of Simpson's paradox.

REFERENCES

D. C. Baldus and J. W. L. Cole. 1980. *Statistical Proof of Discrimination.* Colorado Springs, Colo.: Shepard's. [The definitive reference.]

Peter J. Bickel, Eugene A. Hammel, and J. William O'Connell. 1975. "Sex Bias in Graduate Admissions: Data from Berkeley." *Science* 187:398–404. [Reprinted, with an interesting discussion between William Kruskal and Peter Bickel, in *Statistics and Public Policy,* W. B. Fairley and F. Mosteller, eds. 1977. Reading, Mass.: Addison-Wesley, pp. 113–130.]

Paul Meier, Jerome Sacks, and Sandy L. Zabell. 1986. "What Happened in *Hazelwood*: Employment Discrimination, Statistical Proof, and the 80% Rule." In *Statistics and Law,* M. DeGroot, S. Fienberg, and J. Kadane, eds. 1986. New York: Wiley, pp. 1–40. [Discusses some of the issues raised in this article in detail. The entire volume is a valuable source of information about recent legal uses of statistics.]

Frederick Mosteller, Robert E. K. Rourke, and George B. Thomas, Jr. 1970. *Probability with Statistical Applications,* 2nd ed. Reading, Mass.: Addison-Wesley. [An exceptionally lucid textbook with many examples.]

Clifford H. Wagner. 1982. "Simpson's Paradox in Real Life." *The American Statistician* 36:46–48. [Discusses several real-life instances of Simpson's paradox.]

Hans Zeisel. 1969. "Dr. Spock and the Case of the Vanishing Women Jurors." *The University of Chicago Law Review* 37:1–18. [A classic jury discrimination case.]

Statistics in Jury Selection:

How to Avoid Unfavorable Jurors

S. James Press *JurEcon, Inc., Los Angeles and University of California, Riverside*

Statistical methodology can be used in a jury trial to help attorneys to select out those jurors who are likely to be unfavorable. I will first trace the origin of the jury system and comment on various problems associated with picking a jury at random; then I will discuss peremptory challenges and the use of statistics to make those challenges most effective in avoiding jurors who are likely not to favor the client.

Colonists of English origin brought with them to the American continent legal procedures with which they were familiar, including the jury system. In 1606, James I of England granted a charter to the Virginia Company, including the right to "trial by jury." The right was later granted to Rhode Island (1663), New York (1664), New Jersey (1677), and Maryland (1693). Ultimately, the right to trial by jury was guaranteed to all Americans by the Bill of Rights, in both criminal trials (the Sixth Amendment), and civil trials (the Seventh Amendment). But the Constitution was vague about the "number of persons" who should make up a jury, the "kinds of persons" who should be on a jury, and the "way in which the jurors should be selected."

Over the years, various jurisdictions have experimented with juries comprised of various numbers of persons, ranging from six to twelve, with varying degrees of success. (Six persons are typically used in federal jury trials.) Also, two-thirds or three-fourths majority rules have been tried instead of unanimity being required for a jury decision.

In terms of the "kinds of persons" appropriate to serve on a particular jury, how should the notion of "a person has the right to be tried by a jury of his peers" be interpreted? Who are a person's peers? If a person on trial is Black, for example, should the jury contain the same proportion of Blacks as in the venue? In the county? In the state? In the country? The issue is resolved differently in different jurisdictions in the country.

Experts distinguish various stages of the jury selection process. They start with random selection of potential jurors from voter registration lists and lists of automobile owners, using constraints such as "a potential juror should not be required to travel more than 30 miles from home" and dealing with overlapping boundaries of courthouse jurisdictions. Is such a constrained selection process really random? Another stage of selection of jurors involves the elimination of constitutionally unqualified persons. Then there is the elimination of persons considered undesirable by the attorneys (*voir dire*). This step results in the selection of a complete jury (usually 12 persons for a municipal, county, or state jury, or 6 persons for a federal jury), with alternates, from the pool of potential jurors.

Statistical procedures have frequently been used to study various aspects of this selection process. For example, statistical procedures have been used to study how "representative" a particular jury is of the area in which a crime may have been committed—have certain minority groups been systematically excluded? Statistical procedures have also been used to examine how random the selection procedure is. In order to keep this article brief, however, I will confine the discussion to the use of statistics in the voir dire process.

The term voir dire literally means "to speak the truth" (from the French). In practice, a large pool of randomly selected potential jurors is developed (up to 100 persons) in order to select 6 to 12, plus alternates. This pool contains persons who are statutorily eligible to serve. But, as we all know, people have all kinds of prejudices that they bring to a particular case. Ideally, society would like jurors to be impartial in the case they try. For this reason, attorneys for both sides are usually permitted to ask potential jurors a battery of questions designed to identify persons who are likely to be partial. (In federal cases, the questions are generally asked by the judge, although attorneys are usually permitted to present the judge with a list of questions, which he or she may or may not ask.) The attorneys are preassigned a fixed number of *peremptory challenges* (that is, opportunities to excuse a potential juror without having to give any explanation). The attorneys use these peremptory challenges to exclude from the jury persons who are likely to be unfavorable toward their client in the case. The process of interrogating potential jurors in this way, where the potential jurors are asked "to speak the truth," is called the voir dire.

How does an attorney know which potential jurors to challenge? Historically, attorneys challenge jurors on the basis of the attorneys' own experience and intuition. If their visceral reaction to a potential juror is negative (or if he or she seems overtly unfavorable toward the client), the lawyers excuse that potential juror from serving. Today the procedure is sometimes more scientific; it is sometimes based upon statistical analysis and social science.

APPLYING STATISTICAL METHODS

The use of statistics and social science to assist in the voir dire process, in a formal way, dates back to the early 1970s: the *Mitchell–Stans* trial, the *Harrisburg Seven* trial, the *Attica* trials, and others. In these cases scientific methods were employed by the defense (rather than by the prosecution) to assist in the voir dire process (and also to assist in other aspects of the case). The cases typically involved political issues, multiple defendents, and substantial publicity. When there was substantial pretrial publicity, which was often unfavorable to the defense and made it more difficult for the defense to impanel jurors who would not be unfavorable, the defense was generally given additional peremptory challenges as "compensation." Statistical methodology for scientific jury selection was proposed by various research workers. (Their treatments of jury selection bear on a variety of different aspects of the cases, ranging from statistical, social, psychological, and ethical, to legal—see references.)

I will illustrate the methodology with an example that uses the demographic characteristics of potential jurors to improve the chances of not getting unfavorable jurors on a jury. The situation described, the company mentioned, and the data used are all fictitious, but the case has its roots in real disagreements among actual organizations; the actual facts are proprietary. Jury selection methodology was used successfully in these cases, and in many other cases.

The Mandeville Chemical Company (MCC) has been in business for many years manufacturing items containing chemicals. While these chemicals have had many beneficial uses in industrial applications, one of them has been found to be toxic to human beings and can cause a variety of diseases, including cancer. Many workers brought suit against MCC to cover their medical expenses, and MCC paid off. But MCC, in turn, filed suit against its insurance companies for failing to compensate MCC. The insurance companies argued that MCC had never informed them it was in the business of making a dangerous product. Had they known, they would not have agreed to insure. The case was clearly very complicated. It was not a simple case of some "little guy" suing a big corporation; it was a case of "the big" suing "the big." What kinds of jurors would be likely to be unfavorable (and favorable) toward MCC?

The company decided to do a telephone survey of 800 people. These people were to be asked whether they were U.S. citizens, whether they were located within the trial venue (that is, in the area where the case was scheduled to be tried), and some other questions to determine whether they were potential jurors in the case. (Of those surveyed, 720 were found to be potential jurors.) Potential jurors were then asked a battery of questions. Random digit dialing was used to include people with unlisted numbers. In this procedure the first three digits of a telephone number are selected to correspond to the geographic area of interest. Then the last four digits are selected at random. The result may be a listed or an unlisted number. This is done repeatedly in that area so that many people are reached. The first three digits are then changed and the process is repeated. With a sample as large as 720, conclusions drawn from the sample about how potential jurors view the case could be generalized, with a high degree of confidence, to the larger population of potential jurors contained in the entire venue. (This generalization can be made when the sample is drawn randomly from the population of all potential jurors in the venue.) Respondents to the survey were informed about the facts in the case and were then asked to give their background characteristics, such as gender, age, ethnicity, income category, and so forth. Some subsidiary questions were asked as well, to determine whether there were any points about the case difficult to understand. (If so, the case for the plaintiff could be honed to clarify such points.) Finally, the respondents were asked how they would vote if they were on a jury trying the case.

INTERPRETING THE RESULTS

It was found that 65% of the 720 persons surveyed who were eligible to be jurors in the trial would be unfavorable toward MCC, and 35% would be favorable. The problem was to determine how to distinguish between the two groups so that socioeconomic "marker" variables for partiality against MCC could be used to determine which jurors should be challenged peremptorily and excused.

Cross tabulations of the fractions of respondents unfavorable and favorable toward MCC (the plaintiff) are shown, for age and for gender, in Tables 1 and 2, respectively.

When MCC first looked at the 35% of the total number of respondents who favored the plaintiff, it didn't know anything about them. Examination of Table 1, however, showed that 66% of the population favorable toward MCC was aged 21 to 40 (0.23/0.35), with only 11% aged 41 to 55 (0.04/0.35), 6% aged 56 to 70 (0.02/0.35), and 17% older than 70. More important, MCC then

Table 1 Cross tabulation by age

Age	*Absolute Fraction Unfavorable Toward MCC*	*Absolute Fraction Favorable Toward MCC*	*Totals*	*Fraction of Age Group Favorable Toward MCC*
21–40	0.37	0.23	0.60	0.38
41–55	0.10	0.04	0.14	0.29
56–70	0.03	0.02	0.05	0.40
Above 70	0.15	0.06	0.21	0.29
Totals	0.65	0.35	1.00	0.35

Table 2 Cross tabulation by gender

Gender	*Absolute Fraction Unfavorable Toward MCC*	*Absolute Fraction Favorable Toward MCC*	*Totals*	*Fraction of Gender Group Favorable Toward MCC*
Male	0.52	0.08	0.60	0.13
Female	0.13	0.27	0.40	0.68
Totals	0.65	0.35	1.00	0.35

knew that member of the 20 to 40 group were more likely to be favorable toward the company than were older persons, although the few in the 56 to 70 category were slightly more likely to be favorable (0.40 versus 0.38). It is also clear from Table 1 that the 21 to 40 group was more likely to be favorable toward MCC than was the 41 to 55 group (0.38 versus 0.29). Table 2 shows that 13% of males were likely to be favorable toward MCC (0.08/0.60), and that females were five times as likely as males to be favorable toward MCC (0.68/0.13). Thus MCC knew that if it were to do its best through peremptory challenges to eliminate potentially unfavorable jurors, it should try to avoid older persons and males. Other variables could be used as well. Although these numbers might have been found for this particular case, in other cases totally different sets of numbers are likely to emerge. That is, people's demographic characteristics tend to shape their views according to the context of the case.

Since information about ethnicity, education, income, blue collar versus white collar, and other variables was also available, cross tabulations of these other variables could also have been carried out. For example, Table 1 could be expanded into a higher dimensional table (that is, one containing more variables) so that we could record the fraction of persons who are, say, simultaneously male, college graduates, white, and so on. More-informed conclusions could then be drawn. Also, statistical models could be built to understand the types of people who were not favorable toward the plaintiff.

Of course background characteristics are often only very crude markers of juror voting behavior. Better markers are variables that are surrogates for attitudes, actual behavior, political and social beliefs, and so forth. Such variables can be measured by using a telephone survey to ask such questions as which newspapers the respondent reads, what is the respondent's political affiliation, religious preference, to which clubs or organizations does the respondent belong, what was the respondent's major in college, and whether he or she went to college. Social science research has shown that such variables are likely to be better indicators of juror voting behavior than demographic variables.

The task of an attorney is to represent the client as well as possible. That is, the attorney is an advocate of the client's position. The attorney uses jury selection methods to further the client's case. If this methodology minimizes the chances that some jurors will be a priori hostile to the client's case, regardless of the evidence, the overall justice system benefits.

PROBLEMS

1. Which of the four age groups in Table 1 had the largest number of people unfavorable toward MCC?

2. Which age group in Table 1 had the smallest fraction of people unfavorable toward MCC?

3. Suppose that data are jointly available for age (old versus young) and gender (male versus female) instead of separately as in the text and that the joint information is distributed as follows:

	Absolute Fraction Unfavorable Toward MCC			*Absolute Fraction Favorable Toward MCC*			*Totals*
	Male	*Female*	*Total*	*Male*	*Female*	*Total*	
Young	.27	.10	.37	.06	.17	.23	.60
Old	.25	.03	.28	.02	.10	.12	.40
Totals	.52	.13	.65	.08	.27	.35	1.00

   a. In each of the four age-gender groups find the fraction favorable toward MCC.
   b. Are any of the fractions favorable toward MCC you found in Problem 3a higher than the highest of the four in Table 1 and the two in Table 2? On this basis, which two age-gender groups will the attorneys for MCC most tend to challenge and which two will the defense most tend to challenge?

4. a. In the table for Problem 3, suppose the subtotals remain fixed and you are allowed to decrease the number in the unfavorable male-young cell. What is the smallest value it could have?
   b. Find the smallest entry possible in the favorable male-old cell, keeping the subtotals fixed.

REFERENCES

Reid Hastie, Steven D. Penrod, and Nancy Pennington. 1983. *Inside the Jury.* Cambridge, Mass.: Harvard University Press.

David Kairys, ed. 1975. *The Jury System: New Methods for Reducing Prejudice.* Philadelphia: Philadelphia Resistance Print Shop.

James W. Loewen. 1982. *Social Science in the Courtroom.* Lexington, Mass.: D. C. Heath.

W. McCart. 1965. *Trial by Jury: A Complete Guide to the Jury System,* 2nd ed. Philadelphia: Chilton Books.

S. J. Press. 1988. *Statistics in Jury Selection.* Department of Statistics Technical Report No. 163. Riverside, Calif.: University of California.

J. Schulman, P. Shiver, R. Colman, B. Enrich, and R. Christie. 1973. "Recipe for a Jury." *Psychology Today* 6(May): 37–44, 77–84.

Jon M. Van Dyke. 1977. *Jury Selection Procedures.* Cambridge, Mass.: Ballinger.

Measuring the Effects of Social Innovations by Means of Time Series*

Donald T. Campbell *Lehigh University*

**Supported in part by NSF grant GS 1309X.*

We live in an age of social reforms, of large-scale efforts to correct specific social problems. In the past most such efforts have not been adequately evaluated: usually there has been no scientifically valid evidence as to whether the problem was alleviated or not. Since there are always a variety of proposed solutions for any one problem, as well as numerous other problems calling for funds and attention, it becomes important that society be able to learn how effective any specific innovation has been.

From the statistician's point of view, the best designed experiments, whether in the laboratory or out in the community, involve setting up an *experimental group* and a *control group* similar in every way possible to the experimental group except that it does not receive the same experimental treatment. The statistician's way of achieving this all-purpose equivalence of experimental and control groups is randomization. Persons (or plots of land or other units) are assigned at random (as by the roll of dice) to either an experimental or a control group. After the treatment, the two groups are compared, and the differences that are larger than chance would explain are attributed to the experimental treatment. This ideal procedure is beginning to be used in pilot tests of social policy, as in the New Jersey negative income-tax experiment (Kershaw and Fair, 1976; Watts and Rees, 1977) where several hundred low-income employed families who agreed to cooperate were randomly assigned to experimental groups (which received income supplements of differing sizes) and a control group (which received no financial aid). The effects of this aid on the amounts of other earnings, on health, family stability, and the like were then studied. (Joseph Newhouse's essay in this book describing an experiment in health insurance illustrates this approach.)

Unfortunately, while such experimental designs are ideal, they are not often feasible. They are impossible to use, for example, in evaluating any new program that is applied to all citizens at once, as most legal changes are. In these more common situations much less satisfactory modes of experimental inference must suffice. The *interrupted time series design,* on which this paper will concentrate, is one of the most useful of these quasi-experimental designs. The proper interpretation of such data presents complex statistical problems, some of which are not yet adequately solved. This essay will touch upon a number of these problems, in terms of words and graphs rather than mathematical symbols. The discussion will start by considering two actual cases.

THE CONNECTICUT CRACKDOWN ON SPEEDING

On December 23, 1955, Connecticut instituted an exceptionally severe and prolonged crackdown on speeding. Like most public reporting of program effectiveness, the results were reported in terms of simple before-and-after measures:

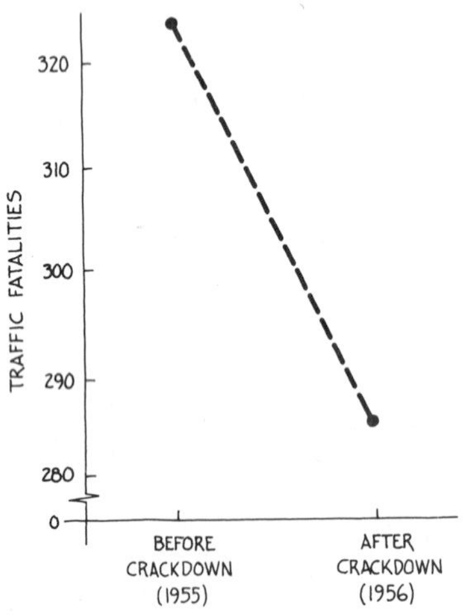

Figure 1 *Connecticut traffic fatalities, 1955–56. Source: Campbell and Ross &1968)*

a comparison of this year's figures with those of the year before. That is, the 1956 total of 284 traffic deaths was compared with the 1955 total of 324, and the governor stated, "With a saving of 40 lives in 1956 . . . we can say the program is definitely worthwhile." Figure 1 presents the data graphically. But this simple quasi-experimental design is very weak and deceptive. There are so many other possible explanations for the change from 324 to 284 highway fatalities. In attributing all of this change to his crackdown, the governor is making an implicit assumption that without the crackdown there would have been no change at all. A time series presentation, using the fatality records of several prior and subsequent years, adds greatly to the strength of the analysis. Figure 2 shows such data for the Connecticut crackdown. In this larger context the 1955–1956 drop looks trivial. We can see that the implicit assumption underlying the governor's statement was almost certainly wrong.

To explore this more fully, turn to Figure 3, which presents in a stylized manner how an identical shift in values before and after a treatment can in some instances be clear-cut evidence of an effect and in others be no evidence at all of a change. Thus with only 1955 and 1956 data to go on, that drop of 40 traffic fatalities shown in Figure 1 might have been a part of a steady annual drop already in progress (the reverse of the steady rise in line F of Figure 3), or of an unstable zigzag (as in line G of Figure 3). Figure 2 shows that in Connecticut the unstable zigzag is the case. The 1955–1956 drop is about the same size as the drops of 1951–1952, 1953–1954, and 1957–1958, times when no crackdowns were present to explain them. Furthermore, the 1955–1956 drop is only half the size of the 1954–1955 rise. Thus with all this previous instability in full graphic view, one would be unlikely to claim all of any year-to-year change as due to a crackdown, as the governor seemed to do.

Later we shall examine Figure 2 again to raise a more difficult problem of inference. But before we do this, let's spend more time on the stability issue, with the help of an illustration from a reform that even a skeptical methodologist can believe was successful.

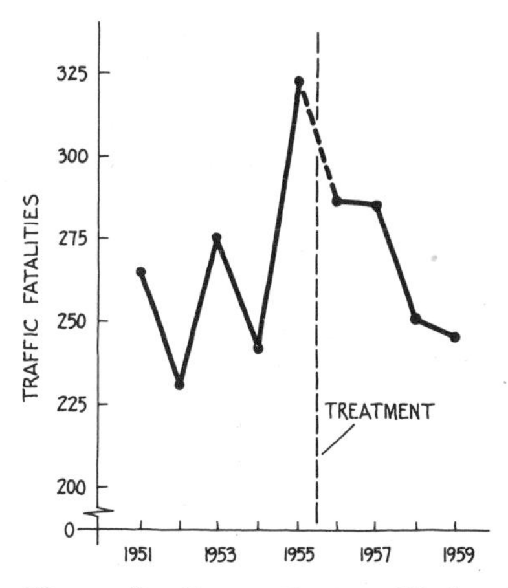

Figure 2 *Connecticut traffic fatalities, 1951–59. Source: Campbell and Ross (1968)*

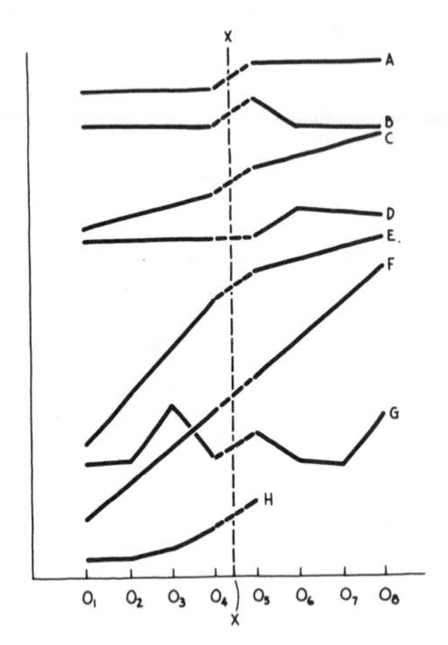

Figure 3 *Some possible outcome patterns from the introduction of a treatment at point X into a time series of measurements, 0_1 to 0_8. The 0_4 to 0_5 gain is the same for all time series, except for D, while the legitimacy of inferring an effect varies widely, being strongest in A and B, and totally unjustified in F, G, and H. Source: Campbell and Stanley (1963)*

THE BRITISH BREATHALYSER CRACKDOWN OF 1967

In September 1967 the British government started a new program of enforcement with regard to drunken driving. It took its popular name from a device for ascertaining the degree of intoxication from a sample of a person's breath. Police administered this simple test to drivers stopped on suspicion, and if it showed intoxication, then took them into the police station for more thorough tests. This new testing procedure was accompanied by more stringent punishment, including suspension of license. Figure 4 shows the effect of this crackdown on Friday and Saturday night casualties (fatalities plus serious injuries). The effect is dramatically clear. There is an immediate drop of around 40% and a leveling off at a level that seems some 30% below the precrackdown rate,

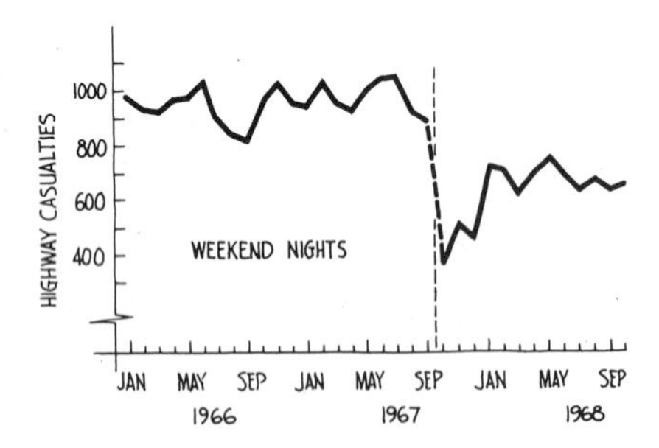

Figure 4 *Effects of the September 1967 English Breathalyser crackdown on drunken driving. Fatalities plus serious injuries, Fridays and Saturdays, 10:00 P.M. to 4:00 A.M., by month. Source: Ross, Campbell, and Glass (1970)*

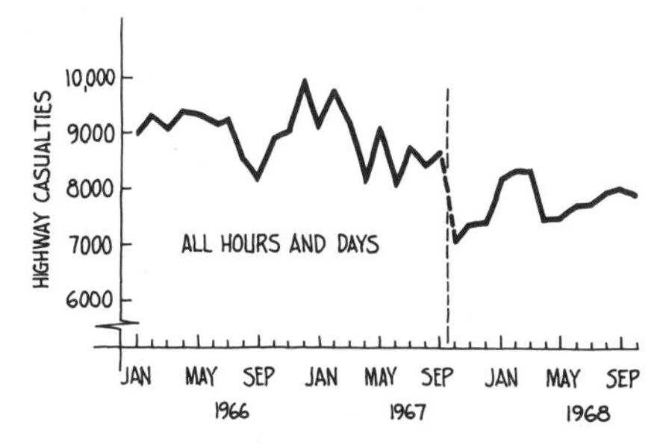

Figure 5 *Effects of the September 1967 English Breathalyser crackdown. Fatalities plus serious injuries, all hours and days, by month. Source: Ross, Campbell, and Glass (1970)*

although this is hard to tell for sure since we don't know what changes time would have brought in the casualty rate without the crackdown.

Does the effect show up when casualties at all hours of all days are totaled? Figure 5 shows such data. While the effect is probably still there, it is certainly less clear, the crackdown drop being not much larger than the unexplained instability of other time periods. (The crackdown drop is, however, the largest month-to-month change, not only during the plotted period but also for a longer period going back to 1961, for data from which the seasonal fluctuations have been removed.)

THE STATISTICAL ANALYSIS OF INSTABILITY

The problem of the statistician is to formalize the grounds for inference that we have used informally or intuitively in our judgments from these graphs. It is clear that the more unstable the line is before the policy change or treatment point, the bigger the difference has to be to impress us as a real effect. One approach of statisticians is to assume that the time series is a result of a general trend plus specific random deviations at each time period. The theory of this type of analysis is well worked out when the random deviations at each point are completely independent of deviations at other points. But in real-life situations the sources of deviation or perturbation at any one point are apt to be similar for adjacent and near points in time, and dissimilar for more remote points. This creates deceptive situations both for statistical tests of significance and for visual interpretation. Figures 6 and 7 illustrate this with a computer-simulated time series. For each point in time, times 1 to 40, there is a *true score*. These true scores, if plotted, would make a straight diagonal line from a lower left score of 0 to an upper right score of 40, with no bump whatsoever at the hypothetical treatment point. These true scores are the same for Figures 6 and 7. To each true score a randomly chosen deviation has been added or subtracted. In Figure 6 the deviation at each time point is drawn independently of every other deviation. This is a simulation of the case in which the hypothetical treatment introduced between time periods 20 and 21 has no effect at all. Occa-

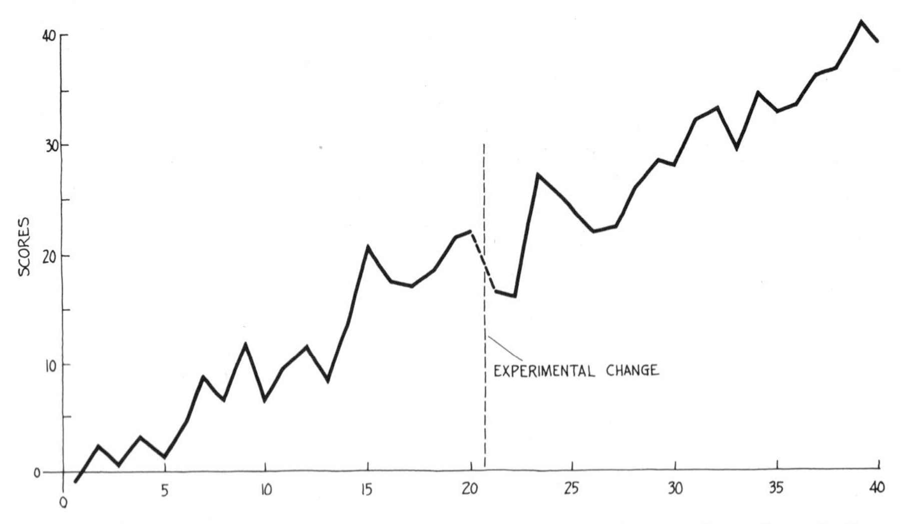

Figure 6 *Simulated time series with independent error. Source: Ross, Campbell, and Glass (1970)*

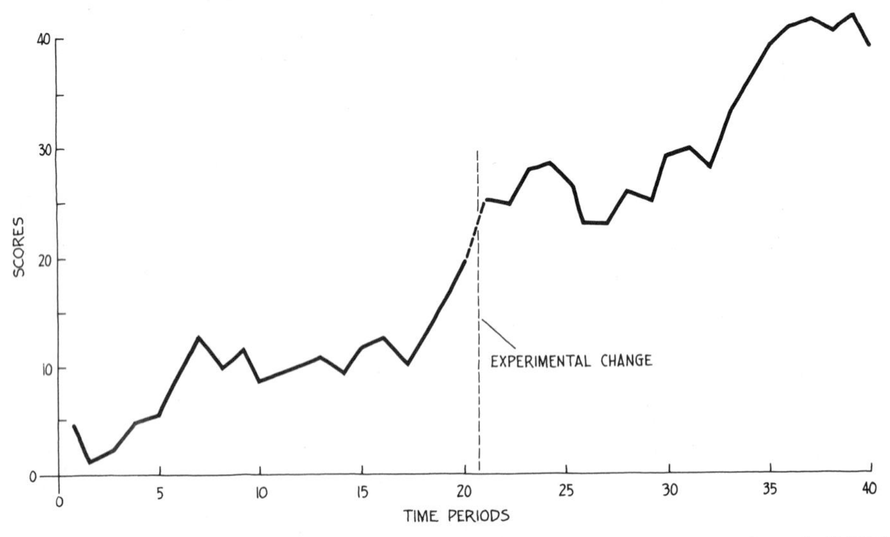

Figure 7 *Simulated time series with correlated (lagged) error. Source: Campbell (1969)*

sionally, by chance, random deviations will occur in such a pattern so as to make it look as though the treatment had an effect, as perhaps in Figure 6. It is the task of *tests of significance* to estimate when the difference from before treatment to after treatment is more than such random deviations could account for. Statistical formulas have been worked out that do this well in the case of independent deviations such as Figure 6 illustrates.

Figure 7 is based upon the same straight diagonal line as Figure 6. It has the same magnitude of deviations added. But the deviations are no longer independent. Instead, four smaller deviations have been added at each point, in a staggered or lagged pattern. A new deviation is introduced at each time period and persists for three subsequent periods. As a result, each point shares three such deviations with the period immediately prior and three with the period immediately following. It shares two deviations with periods two steps away in either direction, and one deviation with periods three steps away. For periods four or more steps away, the deviations are independent. While Figure 7, like Figure 6, is a straight line distorted by random error, note how much more dynamic and cyclical it seems. Such nonindependent deviations mislead both visual judgments of effect and tests of significance that assume independence, through producing judgments of statistically significant effect much too frequently. (To emphasize the lack of any true or systematic departures, let it be emphasized that were one to repeat each simulation 1,000 times, and to average the results, each average would approximate the perfectly straight diagonal line of the true scores. That is, Figures 6 and 7 impose an underlying linearity, which will usually be inappropriate.) There are a variety of ways in which statisticians are attempting to get appropriate tests of significance for the real-life time series in which nonlinear general trends and nonindependent deviations are characteristic. (Probably most appropriate are the procedures of Box and Tiao, 1975; Cook and Campbell, 1979; McCleary and Hay, 1980.)

REGRESSION ARTIFACTS

We have moved from simple problems of inference to more complex ones. We will return soon to some more easily understood problems. But before doing that, let us attempt to understand a final difficult problem, known in one statistical tradition as *regression artifacts.* If we can be sure that the policy change took place independently of the ups and downs of the previous time periods, there is no worry. But if the timing of the policy change was chosen just because of an extreme value immediately prior, then a regression artifact will be sufficient to explain the occurrence of subsequent less extreme values. To see if a regression artifact might be at work in the Connecticut case, let us return to Figure 2. Here we can note that the most dramatic change in the whole series is the 1954–1955 increase. By studying the newspapers and the governor's pronouncements, we can tell that it was this striking increase that caused him to initiate the crackdown. Thus the treatment came when it did because of the 1955 high point.

In any unstable time series, after any point that is an extreme departure from the general trend, the subsequent points will on the average be nearer the general trend. Try this out on Figure 6. Move your eye from left to right, noting each point that is "the highest so far." For most of these, the next point is lower, or has *regressed* toward the general trend. Such regression subsequent to points selected for their extremity is an automatic feature of the very fact of instability and should not be given a causal interpretation. Applied to Figure 2, this means that even with no true effect from the crackdown at all, we would expect 1956 to be lower than the extreme of 1955.

OTHER REASONS FOR SHIFTS IN TIME SERIES

It is going to be important for administrators, legislators, the voting public, and other groups of nonstatisticians to be able to draw conclusions from time series data on important public programs. For this reason, two further points will be made that are less directly statistical. First note that there are many reasons for abrupt shifts in time series other than the introduction of a program change. One very deceptive reason is a shift in recordkeeping procedures. Such shifts are apt to be made at the same time as other policy changes. For example, a major change in Chicago's police system came in 1959 when Professor Orlando Wilson was brought in from the University of California to reform a corrupt police department. Figure 8 shows his apparent effect on thefts—a dramatic *increase.* This turns out to be due to his reform of the recordkeeping system, and the rise was anticipated for that reason.

In a real situation, unlike in an insulated laboratory, many other causes may be operating at the same time as the experimental policy change. Thus in Connecticut or in England, a drop in traffic casualties might have been due to especially dry weather, or fewer cars on the road, or to new safety devices, or to a multitude of other factors. If one had been able to design an experiment with the randomized control groups discussed at the beginning of this essay, such explanations would have been ruled out statistically. However, setting up such an experiment would have been impossible in these two situations. We must instead try to rule out these rival explanations of an effect in other ways. One useful approach is to look at newspaper records of rainfall, changes in traffic density, and other possible causes of the shift.

Another approach is to look for some control comparison that should show the effects of these other causes, if they are operating, but where the specific reform treatment was not applied. For Connecticut, the data from four nearby states are relevant, as shown in Figure 9. All of these states should have been affected by changes in weather, new safety features in cars, and so forth. While these data support the notion that the 1956 Connecticut fatalities would have been lower than 1955 even without the crackdown, the persisting decline throughout 1957, 1958, and 1959 is steepest in Connecticut and may well indicate a genuine effect of the prolonged crackdown. While visually we have little

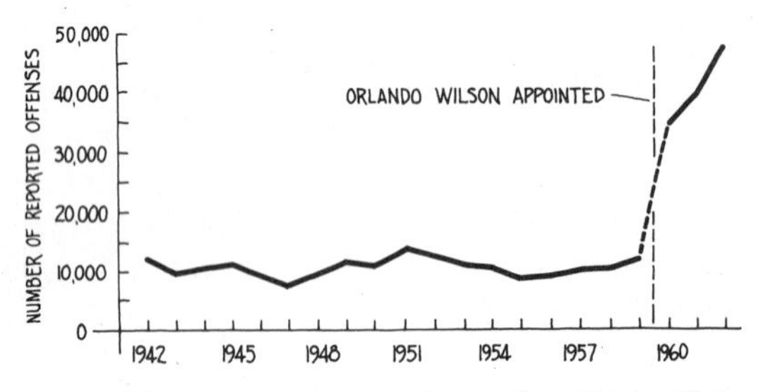

Figure 8 *Reported larcenies under $50 in Chicago from 1942–1962. Source: Campbell (1969). Data from Uniform Crime Reports for the United States, 1942–62.*

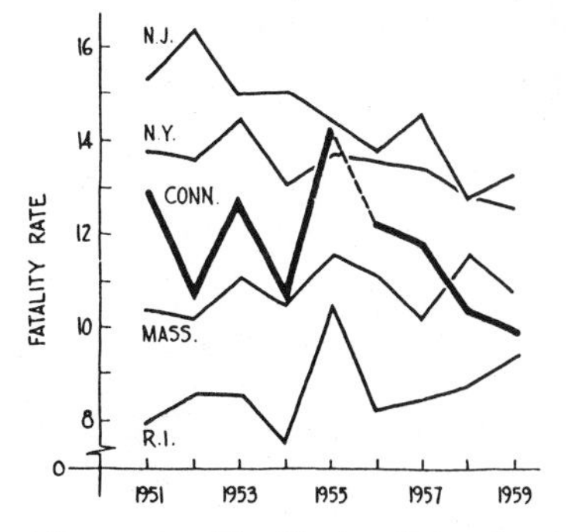

Figure 9 *Traffic fatalities for Connecticut, New York, New Jersey, Rhode Island, and Massachusetts (per 100,000 persons). Source: Campbell (1969)*

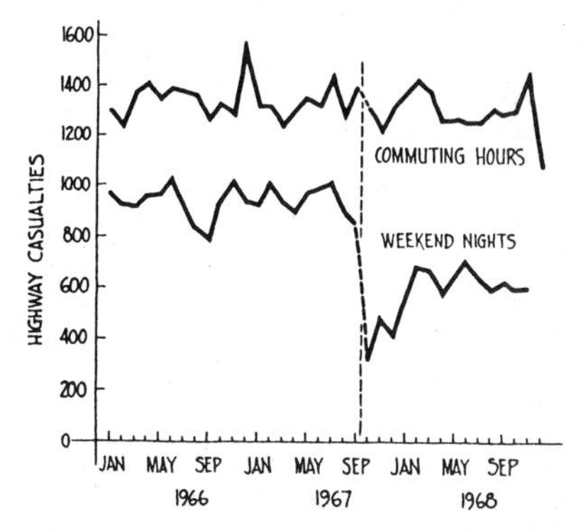

Figure 10 *A comparison of casualties during closed hours (commuting hours) and weekend nights in the English Breathalyser crackdown. Source: Ross, Campbell, and Glass (1970)*

difficulty in using these supplementary data, the statistician has many real problems in combining them all in an appropriate test of significance.

For England, there were no appropriate comparison nations available. But British pubs are closed before and during commuting hours, so casualties from such hours provide a kind of comparison base, as shown in Figure 10. Unfortunately there is a lot of instability in these data so they do not enable us to estimate with much confidence the degree to which the initial crackdown effects are persisting.

A FINAL NOTE

There are several general lessons to be learned from these brief illustrations. The first involves the distinction between *true experiments,* in which experimental and control groups are assigned by randomization, and *quasi-experiments.* True experiments, when they are possible, offer much greater power and precision of inference than do quasi-experiments. The administrators of social innovations, in consultation with statisticians, should attempt to use such designs where possible. Where true experiments are not possible, or have not been used, there are some quasi-experimental designs, such as the interrupted time series, that can be very useful in evaluating policy changes. These too require statistical skill to avoid misleading conclusions. Evaluation of social innovations is an important and challenging area of application for modern statistics.

PROBLEMS

1. Why is the design on which this essay concentrates called an *interrupted* time series design?
2. How does Figure 2 change your perception of Figure 1?
3. Consider Figure 3. Suppose that X is a currency devaluation that results in a price increase between 0_4 and 0_5. What effect will X have on the price of Product A? Product C? Product F?
4. Consider Figure 3. Why does the author say that one can most legitimately infer an effect in A and B while one would be totally unjustified in inferring one in F, G, and H?
5. Consider Figures 4 and 5. Is the effect of the Breathalyser crackdown more pronounced in one of the figures? If so, which one?
6. a. Explain the difference between independent and lagged error.
   b. How would you characterize the effects of the two types of error on the plots of data in Figures 6 and 7?
7. What does the author mean when he says Figure 7 is more "dynamic and cyclical" than Figure 6?
8. Explain what the author means by the term *regression artifact.*
9. Refer to Figure 8. Explain the sharp increase in the number of reported offenses when Orlando Wilson was appointed.
10. Should the 1955–1956 traffic fatality decline in Rhode Island (Figure 9) be attributed to the speeding crackdown in neighboring Connecticut? Explain your answer.
11. What is a *true experiment*? What is a *quasi-experiment*?
12. Comment on this statement: Quasi-experiments are used in evaluating policy changes because it is virtually impossible to apply true experimental design to social situations.

REFERENCES

G. E. P. Box and G. C. Tiao. 1975. "Intervention Analysis with Applications to Economic and Environmental Problems." *Journal of the American Statistical Association* 70: 70–79.

D. T. Campbell. 1969. "Reforms as Experiments." *American Psychologist* 24(4): 409–429.

D. T. Campbell and H. L. Ross. 1968. "The Connecticut Crackdown on Speeding: Time-Series Data in Quasi-Experimental Analysis." *Law & Society Review* 3(1): 33–53.

D. T. Campbell and J. C. Stanley. 1963. "Experimental and Quasi-Experimental Designs for Research on Teaching." In *Handbook of Research on Teaching,* N. L. Gage, ed. Chicago: Rand-McNally, pp. 171–246. (Reprinted as *Experimental and Quasi-Experimental Designs for Research.* 1966. Chicago: Rand-McNally.)

T. D. Cook and D. T. Campbell. 1979. *Quasi-Experimentation: Design and Analysis Issues for Field Settings.* Chicago: Rand-McNally.

D. Kershaw and J. Fair, eds. 1976. "Operations, surveys and administration," Vol. 1: *The New Jersey Income-Maintenance Experiment.* New York: Academic Press.

R. McCleary and R. A. Hay, Jr. 1980. *Applied Time Series Analysis for the Social Sciences.* Beverly Hills: Sage.

H. L. Ross, D. T. Campbell, and G. V. Glass. 1970. "Determining the Social Effects of a Legal Reform: The British 'Breathalyser' Crackdown of 1967." *American Behavioral Scientist* 15(1): 110–113.

H. W. Watts and A. Rees, eds. 1977. "Expenditures, health, and social behavior; and the quality of the evidence," Vol. 2 and 3: *The New Jersey Income-Maintenance Experiment.* New York: Academic Press.

Election Night on Television

Richard F. Link *Richard F. Link & Associates, Inc.*

During the evening of the first Tuesday in November in even-numbered years, millions of people all over the United States watch the election shows provided by the three major networks. The viewers see a rapid tabulation of the votes cast for the major state offices of senator and governor, and in years when a president is elected, a rapid tabulation of the presidential vote by state and for the nation. They also see a tabulation of the votes for members of the House of Representatives. They usually hear an announcement of the winner after only a small percent of the vote has been reported, often within minutes of the closing of the polls. As the evening progresses they are treated to analyses that explain how given candidates won, that is, where their strength and weakness lay and why it appeared that they won.

Massive machinery operates behind this effort. This machinery is physical in the sense that it requires a very elaborate communications network and extensive use of computers, but it is also statistical and mathematical in the sense that it requires rapid summaries and interpretations so that the findings can quickly be passed to the viewing public.

I shall not attempt to describe the complete organization necessary to produce the election night show, but shall describe the three parts of the show that lean most heavily upon computer and statistical technology: vote tabulation, projection of winners, and detailed analysis of the vote. The three networks use basically the same vote tabulation system, but they differ in their methods for projecting winners and in their analysis of the vote. I shall describe only the method of projection used by one network (NBC), at least until 1988.

Before discussing the procedures and methods used today, let's look at a brief history of the reporting of election night results to give a feel for why and how today's shows came about.

A BRIEF HISTORY OF ELECTION REPORTING

Persons living in the United States, as in other free societies that hold elections, have always had an intense interest in the outcome of elections. Interest is most intense for the elections that involve the presidency, is reasonably high for gubernatorial and senatorial elections, and at least the numbers of Republicans and Democrats composing the House of Representatives are of concern, even though the election of a particular member usually does not have national significance.

Thus election results have always been news of great interest. Until about 1928, this news reached the public via the newspapers. In general the coverage was relatively slow and incomplete. Radio changed this situation, and election reporting was speeded up. For example, radio reported the upset victory of Harry S. Truman in the early hours following election day in November 1948. Television began to report elections on a national scale in 1952 and has increased its scope and coverage and speed of gathering the vote since then. Extensive coverage of election primaries was introduced during the presidential year of 1964 and continues to be a feature of television reporting today even in "off-years."

Two factors influence speed of coverage: how quickly the vote is obtained from its source (basically a precinct), and then how soon it is reported. The speed of reporting the vote, once collected, was greatly increased by reporting via radio as opposed to reporting via newspaper. This reporting speed has not been particularly increased by television. Both radio and television are capable of essentially instantaneous reporting. The speed of vote collection, however, has been greatly increased by the television networks. It is worth reviewing the collection procedures utilized in the past and today.

The United States has approximately 175,000 precincts. In the official electoral machinery, the precinct vote is usually forwarded to a county collection center, and then to a state center, often to the Secretary of State there, who then certifies the official vote. Final official collection and certification frequently take several months. The precinct vote, however, is forwarded to the county level fairly rapidly, perhaps by phone or courier, and the vote at the county level is often quickly available on an unofficial basis. The job of collecting the vote at the county level is much less arduous than that of collecting at the precinct level since there are only about 3,000 counties in the country. The vote can be collected faster, nonetheless, if it is collected at the precinct level, and this is the basic innovation that television introduced to vote collection. The networks with their large economic resources were instrumental in establishing a mechanism for obtaining the vote at the precinct level and communicating it by phone to a central location where it could be processed by a computer.

Competition by television networks in the area of extensive vote collection became very intense by the primary elections of 1964. That year the New Hampshire primary saw all three of the major television networks collecting and reporting the vote at the precinct level. In fact, some wags have said that there were more television workers in New Hampshire during the 1964 primary than voters, or to put it another way, that it would have been cheaper to bring the New Hampshire voters to New York to vote at a central location than to collect the vote in New Hampshire. Needless to say, these remarks are exaggerated, but they do emphasize the magnitude of the expense involved. The competition in collection became more intense that spring, and culminated in the reporting of the California presidential primary where each of the three networks collected the vote in the more than 30,000 precincts in California. This enormous expense brought only a mixed blessing. The newspaper wire services continued to collect and report the vote in the traditional manner, from complete county returns, so that on the day after the election, after the television networks had reported Goldwater the winner, the newspapers all showed Rockefeller with a substantial lead. The reason for this disparity was that Los Angeles County, with approximately a third of the precincts in the state, did not have a complete county report until too late to meet the newspaper deadlines, and Goldwater ran very strongly in Los Angeles.

This confusion was coupled with another, arising from the fact that each network would report its own vote totals at any instant of time, and, since they were being collected independently, at any given moment their totals were all different. All this led to the formation of an organization called the News Election Service (NES) whose sole purpose is to collect the vote and report it to its members. This service was formed by a cooperative effort of the three television networks (ABC, CBS, NBC) and the two wire services (AP, UPI). The NES releases its figures to its members simultaneously, so that at any instant all networks and news services are able to report the same basic data to the public.

THE NEWS ELECTION SERVICE

The massive NES operation functions in the following manner. Reporters, called stringers, are on duty at more than 100,000 of the largest of the 175,000 precincts in the country and at each of the 3,000 county reporting centers. These reporters collect the vote at the precinct and county levels and then phone the vote to a central location, adding enough information to identify the source of the report. The vote information is then put into a computer that checks its apparent validity; for example, if it is a precinct report, it checks that the vote does not exceed registration in that precinct, or if a county report, that the number of precincts in the county has not been exceeded. Because registration figures and data on number of precincts are not exactly accurate at this time the check depends upon a statistical tolerance rather than an absolute cutoff. Once the report has been checked, if it is a precinct report it is added to the precinct results already reported for that county. If it is a county report, it replaces the previous county report (the county reports are made on a cumulative basis). At regular intervals the computer generates a vote report for each election race for each county in the state and also provides a state total for the presidential, senatorial, and gubernatorial races and a national total for the presidential race. The summary report is generated by comparing the county votes from the county reports with the votes in the county calculated from the precinct reports. It uses the larger figure for the county figure, and then sums over the counties in the state to obtain a state figure. In presidential years an additional summation is made over the states to obtain a national vote figure for the presidency. In addition to providing summary vote totals, percentages for each candidate are reported, as is the fraction of precincts reporting. Similar summaries for precincts are grouped for congressional districts to evaluate races for the House.

Once the information has been calculated, the computer releases the information to its clients. It provides this information in printed form both at the computer location and at the television studio, and it also makes the information available via telephone lines that can be used for input into the various network computer systems.

Extensive preparations must be done before the election not only to train the stringers but also to gather the registration figures for the precincts, to find the number of precincts in each county, and to collect other basic data. This information is essential for the checking process and for accurate reporting of the fraction of the vote that has been counted.

Additional preparation must also go into the operation that gets the vote into the computer and into the programming of the computer to accept the vote, add it properly, and report it correctly. The general name of the operation that prepares the computer to work properly is called *coding*. Its importance cannot be overestimated. In 1968 mistakes made in the instructions for the computer caused the computer to malfunction, and the vast stream of votes from NES dried up to a trickle shortly after midnight (EST) election night. This malfunction was partly responsible for the uncertainty about the winner in the presidential race, which was not reported by the television networks until Wednesday morning.

This brings us to an area that is still the subject of intensive competition among the three networks. Although they are all constrained by their common use of NES totals to report the same vote totals, they are not constrained in the interpretation of these vote totals; for although the vote total at any instant in time may be informative, the real interest in an election lies in who wins, by how much, and why.

PROJECTING ELECTION WINNERS

The rapid collection and reporting of the vote requires a great deal of organization, computer capability, and communication equipment. All that activity, nonetheless, goes simply to adding up the vote. The question for the election forecaster always remains: When can I be reasonably sure I have tabulated enough of the vote to decide who will be the ultimate winner?

An easy answer to that question is: Wait until all the votes are counted—but this may take days. Statistical theory, however, sometimes allows us to give an answer earlier. Sometimes it allows us to determine the winner of an election when only a fraction of 1% of the vote has been reported to the analyst. It happens frequently that projections can be made on the basis of information collected by the network and available to the analysts in the television studio before a single vote has been posted for the television audience because NES has not yet produced vote totals (which, by agreement, are the only ones that can be released to the public).

The projection of election night winners requires a combination of historical information, statistical theory for the construction of an appropriate mathematical model of the vote and for deciding when one is sure enough to make a projection, and the actual election night vote. The networks have different schemes for projecting winners, but all of these schemes have the basic elements we have described. Let's now look at the general scheme used by one network, NBC. We begin with the projection of the winner in a state race, then the projection of the winner in a presidential election, and, finally, the projection of the composition of the House of Representatives.

STATE RACES

The information for projecting the winner of a state race comes from four separate sources. First, a preelection estimate of the percentage each candidate will get is obtained from public opinion polls, newspaper reporters, politicians in the state, and similar sources. This initial estimate, often quite accurate, may give a definite indication of how the race will turn out. Second, interviews of voters as they leave selected voting places provide the Election Day Voter Poll. Respondents fill in the questionnaires themselves so that their responses are not known to the person conducting the poll. Third, the network collects the vote of specially selected precincts, called key precincts, in addition to the vote collected by NES. Typically the network collects votes from 50 to 150 such precincts for each state. Thus a network may have a national precinct collection system utilizing reporters at over 5,000 precincts completely independent of the NES effort. The voting behavior of these key precincts in past elections will already have been carefully analyzed. Fourth, the information from NES is available at a county level.

The information from these four sources is ordered in time. The initial estimate is obviously available first because it comes before election day. The results from the Election Day Voter Poll respondents will be phoned in during the day and processed by computer so that the approximate results for a given race in a given state will be known well before the polls have closed in that state. The vote of the key precincts, which the networks collect themselves, is usually the first actual vote information available to the network. By the agreement forming NES, this key precinct vote information cannot be displayed on the air but it can be used to project winners. If a race is one-sided in a state, the results of the Election Day Voter Poll by itself may be used to call the race as soon as the polls in the state close. If a race is less one-sided but still fairly clear as to its outcome, it may be callable after a few key precincts have reported.

If the race is close, however, more of the special precincts are needed before a winner may be projected, and often it is necessary to use the county information available from NES.

To use the information from the counties, it is necessary to develop a mathematical model. The reason is that different counties have different voting behaviors. For example, New York City is always more Democratic in its vote than the rest of the state of New York. This kind of difference in voting behavior in terms of relative Democratic or Republican leanings can be incorporated into a mathematical model.

The statistical model uses the voting patterns from the recent past. For example, the fraction of the New York State vote in New York City is typically 0.4, and the fraction in the rest of the state 0.6. In a typical past election for governor, the Democratic candidate got 50% of the vote in New York City and 40% of the vote in the rest of the state. His statewide vote, then, was $0.4(50\%) + 0.6(40\%) = 44\%$. Thus New York City was 6% more Democratic than the state average, and the rest of the state was 4% less Democratic than the state average. The fractions can be incorporated into a model so that in this simplest instance, if in a new election the early returns from New York City show 54% for the Democratic candidate and the rest of the state shows 48% for the Democratic candidate, the state projection in percent would be $0.4(54\%) + 0.6(48\%) = 50.4\%$. This indicates that the Democratic candidate would win, although if the returns were very early, this projection would not be considered sufficiently accurate to make an announcement of a victory.

Another factor considered in the statistical model is whether the fraction of the vote assigned to the various parts of the state is accurate for this election. If it snowed heavily in upstate New York, but not in New York City, thus cutting the vote upstate, but not in New York City, the fraction of the vote in the election might be 0.5 for New York City and 0.5 for the rest of the state;

that is, the relative voter turnout in New York City would be higher than normal. In this case the projection would be 0.5(54%) + 0.5(48%) = 51%, indicating a better chance for Democratic victory. Thus the differential turnout must also be considered in the model in order to make vote projections.

The use of computers allows such a model to be constructed using detailed information for all the counties of a state rather than just the two regions in our example, to provide not only a projection but also an indication of the accuracy of the projection, so that one can decide when a projection may safely be announced.

It is useful for the model to include the prior estimate available to the network and results for the special key precincts, so that all information available to the network is effectively utilized. Such a model sometimes allows the results of a race to be called with near certainty, even though only a small fraction of the vote is reported and the race is relatively close.

It is network policy not to predict the winner unless it is almost a certainty that the predicted winner will actually win. The accuracy of the predictions can be gauged by the fact that, during a given evening when over a hundred predictions may be made, there is usually at most one mistake. The use of such models, developed by statistical theory, allows the networks to enforce their policy and at the same time "call" close races because the precision of the estimates developed by the models is always known. One of the most important outputs of the statistical model in this decision problem, as in many others, is the estimated precision of the result.

PRESIDENTIAL RACE

The Election Day Voter Poll revolutionized the way the networks were able to call the results of presidential races. This was the basic device that allowed NBC to announce at 8:15 P.M. (EST) on November 4, 1980 that Ronald Reagan had been elected the fortieth president of the United States. By using the Election Day Voter Poll, NBC was able to ascertain that Reagan would receive 270 electoral votes from states whose voting places had closed by 8:00 P.M. (EST). All the networks made similar early calls in 1984.

Because many people in the West had not yet voted for the president and other candidates at such an early hour (5:00 P.M. P.S.T.), this new technology led to some controversy. Various political solutions have been proposed, such as having a uniform poll closing time across the country. Other solutions that involve news censorship raise severe constitutional problems. Carried to a ridiculous extreme, one might suggest keeping the results secret until the electoral college meets the following January!

When presidential races are not so one-sided, the results may still not be known until the early hours of the morning after election day. In 1968 the NES computers were down for a while, and because this was a very close election, the final result was not known for the many hours it took the networks to gather sufficient information to make a responsible projection of the winner. Even if the Election Day Voter Poll technology had been available in this election, the projection of the winner would have come very late.

HOUSE RACES

A projection of the composition of the House of Representatives requires a model similar to the one used for projection of presidential races, the main difference being that each house seat counts as 1, the prior estimates are ordinarily less reliable than those for states in presidential elections, and the vote is reported only by house district. In addition to the projection and vote information, the networks also provide an *analysis* of the vote. This is the next topic we shall consider.

NEWS ANALYSIS

The Election Day Voter Poll questionnaire has questions about various issues that the candidate may have discussed during the campaign and about the characteristics of the person filling out the questionnaire, such as age, sex, race, religion, income, and so forth, and questions about whom the respondent may have voted for. By using cross tabulations it is possible to analyze the vote in terms of various demographic factors (such as sex, race, and so forth) and various issues. These analyses help explain what the election meant to the electorate. For example, a majority of the electorate might favor freedom of choice even though a candidate who opposed abortion was elected. This means that other factors played a more important role in the candidate's election, and that the candidate cannot correctly claim to have a mandate for his or her position on abortion. Such analyses enable network commentators to flesh out their opinions and qualitative insights with quantitative information, thus offering the viewers a more informed view of the election's context than would otherwise be available.

CONCLUSION

The reporting effort of the television networks represents an area of activity that could not exist without the computer and without modern statistics. It represents a blend of modern technology and the traditional skills of the reporter.

The statistical techniques of vote projection may have other applications. For example, it might be possible by similar methods to establish the pattern of yields of corn county by county in Iowa from historical records, and accurately to estimate the state yield from the yields of only a few early harvesting counties.

PROBLEMS

1. What are the three parts of the election night show that rely most heavily on computer and statistical technology?
2. What are the advantages of precinct level vote collection by the media? The disadvantages?
3. Why was the NES formed? Does this mean that the only data available to the networks are from the NES?
4. Statistical theory enters into winner projection in two ways. Describe them.
5. What are *key precincts*? Are they the same for all networks?
6. A gubernatorial candidate in New York is assured of 60% of the New York City vote and 50% in the rest of the state. As stated in this article, the New York City vote usually represents 40% of the statewide total. What percentage of the total vote can our candidate expect?
7. Our candidate is dismayed. A sudden blizzard has hit New York City on the first Tuesday in November, cutting the city's voter turnout to 30% of the state total. Can the candidate still win?
8. Why is the precision of an estimate important in winner projection?
9. Besides the estimated voting percentages themselves, what is an equally important output of the projection models discussed?

PART THREE

Our Social World

Communicating with Others

People at Work

People at School and Play

Counting People and Their Goods

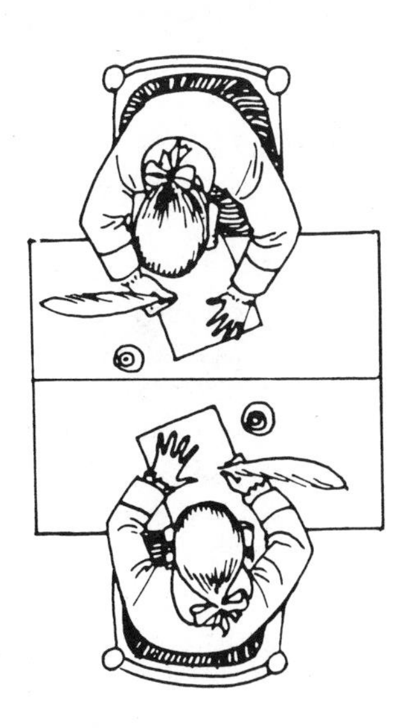

Deciding Authorship

Frederick Mosteller *Harvard University*
David L. Wallace *University of Chicago*

Art, music, literature, and the social, biological, and physical sciences share a common need to classify things: What artist painted the picture? Who composed the piece? Who wrote the document? If paroled, will the prisoner repeat the crime? What disease does the patient have? What trace chemical is damaging the process? In the field of statistics, we call these questions *classification* or *discrimination* problems.

Questions of authorship are frequent and sometimes important. Most people have heard of the Shakespeare-Bacon-Marlowe controversy over who wrote the great plays usually attributed to Shakespeare. A less well known but carefully studied question deals with the authorship of a number of Christian religious writings called the Paulines, some being books in the New Testament: Which ones were written by Paul and which by others? In many authorship questions the solution is easy once we set about counting something systematically. But we treat here an especially difficult problem from American history, the controversy over the authorship of the 12 *Federalist* papers claimed by both Alexander Hamilton and James Madison, and we show how a statistical analysis can contribute to the resolution of historical questions.

The Federalist papers were published anonymously in 1787–1788 by Alexander Hamilton, John Jay, and James Madison to persuade the citizens of the state of New York to ratify the Constitution. Seventy-seven papers appeared as letters in New York newspapers over the pseudonym Publius. Together with eight more essays, they were published in book form in 1788 and have been republished repeatedly both in the United States and abroad. *The Federalist* remains today an important work in political philosophy. It is also the leading source of information for studying the intent of the framers of the Constitution, as, for example, in decisions on congressional reapportionment, since Madison had taken copious notes at the Constitutional Convention.

It was generally known who had written *The Federalist,* but no public assignment of specific papers to authors occurred until 1807, three years after Hamilton's death as a result of his duel with Aaron Burr. Madison made his listing of authors only in 1818 after he had retired from the presidency. A variety of lists with conflicting claims have been disputed for a century and a half. There is general agreement on the authorship of 70 papers—5 by Jay, 14 by Madison, and 51 by Hamilton. Of the remaining 15, 12 are in dispute between Hamilton and Madison, and 3 are joint works to a disputed extent. No doubt the primary reason the dispute exists is that Madison and Hamilton did not hurry to enter their claims. Within a few years after writing the essays, they had become bitter political enemies and each occasionally took positions opposing some of his own *Federalist* writings.

The political content of the essays has never provided convincing evidence for authorship. Since Hamilton and Madison were writing a brief in favor of ratification, they were like lawyers working for a client; they did not need to believe or endorse every argument they put forward favoring the new Constitution. While this does not mean that they would go out of their way to misrepresent their personal positions, it does mean that we cannot argue, "Hamilton wouldn't have said that because he believed otherwise." And, as we have often seen, personal political positions change. Thus the political content of a disputed essay cannot give strong evidence in favor of Hamilton's or of Madison's having written it.

The acceptance of the various claims by historians has tended to change with political climate. Hamilton's claims were favored during the last half of the nineteenth century, Madison's since then. While the thorough historical studies of the historian Douglass Adair over several decades support the Madison claims, the total historical evidence is today not much different from that which historians like the elder Henry Cabot Lodge interpreted as favoring Hamilton. New evidence was needed to obtain definite attributions, and internal statistical stylistic evidence provides one possibility; developing that evidence and the methodology for interpreting it is the heart of our work.

The writings of Hamilton and Madison are difficult to tell apart because both authors were masters of the popular *Spectator* style of writing—complicated and oratorical. To illustrate the difficulty, in 1941 Frederick Williams and Frederick Mosteller counted sentence lengths for the undisputed papers and got averages of 34.5 and 34.6 words for Hamilton and Madison, respectively. For sentence length, a measure used successfully to distinguish other authors, Hamilton and Madison are practically twins.

MARKER WORDS

Although sentence length does measure complexity (and an average of 35 words shows that the material is very complex), sentence length is not sensitive enough to distinguish reliably between authors writing in similar styles. The variables used in several studies of disputed authorship are the rates of occurrence of specific individual words. Our study was stimulated by Adair's discovery—or rediscovery as it turned out—that Madison and Hamilton differ consistently in their choice between the alternative words *while* and *whilst*. In the 14 *Federalist* essays acknowledged to be written by Madison, *while* never occurs whereas *whilst* occurs in 8 of them. *While* occurs in 15 of 48 Hamilton essays, but never a *whilst*. We have here an instance of what are called *markers*—items whose presence provides a strong indication of authorship for one of the men. Thus the presence of *whilst* in 5 of the disputed papers points toward Madison's authorship of those 5.

Markers contribute a lot to discrimination when they can be found, but they also present difficulties. First, *while* or *whilst* occurs in less than half of the papers. They are absent from the other half and hence give no evidence either way. We might hope to surmount this by finding enough different marker words or constructions so that one or more will always be present. A second and more serious difficulty is that from the evidence in 14 essays by Madison, we cannot be sure that he would never use *while*. Other writings of Madison were examined and, indeed, he did lapse on two occasions. The presence of *while* then is a good but not sure indication of Hamilton's authorship; the presence of *whilst* is a better, but still imperfect, indicator of Madison's authorship, for Hamilton too might lapse.

A central task of statistics is making inferences in the presence of uncertainty. Giving up the notion of perfect markers leads us to a statistical problem. We must find evidence, assess its strength, and combine it into a composite conclusion. Although the theoretical and practical problems may be difficult, the opportunity exists to assemble far more compelling evidence than even a few nearly perfect markers could provide.

RATES OF WORD USE

Instead of thinking of a word as a marker whose presence or absence settles the authorship of an essay, we can take the *rate* or *relative frequency* of the use of each word as a measure pointing toward one or the other author. Of course, most words won't help because they were used at about the same rate by both authors. But since we have thousands of words available, some may help. Words form a huge pool of possible discriminators. From a systematic exploration of this pool of words, we found no more pairs like *while-whilst,* but we did find single words used by one author regularly but rarely by the other.

Table 1 shows the behavior of three words: *commonly, innovation,* and *war.* The table summarizes data from 48 political essays known to be written by Hamilton and 50 known to be by Madison. (Some political essays from outside *The Federalist,* but known to be by Hamilton or Madison, have been included in this study to give a broader base for the inference. Not all of Hamilton's later *Federalist* papers have been included. We gathered more papers from outside *The Federalist* for Madison.)

Neither Hamilton nor Madison used *commonly* much, but Hamilton's use is much more frequent than Madison's. The table shows that in 31 of 48 Hamilton papers, the word *commonly* never occurs, but that in the other 17 it occurs one or more times. Madison used it only once in the 50 papers in our study. The papers vary in length from 900 to 3,500 words, with 2,000 about average. Even one occurrence in 900 words is a heavier usage than two occurrences in 3,500 words, so instead of working with the number of occurrences in a paper, we use the rate of occurrence, with 1,000 words as a convenient base. Thus, for example, the paper with the highest rate (1.33 per 1,000 words) for *commonly* is a paper of 1,500 words with two occurrences. *Innovation* behaves similarly, but it is a marker for Madison. For each of these two words, the highest rates are a little over 1 per 1,000.

Table 1 Frequency distributions of rate per 1,000 words in 48 Hamilton and 50 Madison papers for *commonly, innovation,* and *war*

Commonly			*Innovation*			*War*		
Rate per 1,000 Words	*H*	*M*	*Rate per 1,000 Words*	*H*	*M*	*Rate per 1,000 Words*	*H*	*M*
0 (exactly)*	31	49	0 (exactly)*	47	34	0 (exactly)*	23	15
0^+–0.2	Cannot occur†		0^+–0.2	Cannot occur†		0^+–2	16	13
0.2–0.4	3	1	0.2–0.4		6	2–4	4	5
0.4–0.6	6		0.4–0.6	1	6	4–6	2	4
0.6–0.8	3		0.6–0.8		1	6–8	1	3
0.8–1.0	2		0.8–1.0		2	8–10	1	3
1.0–1.2	2		1.0–1.2		1	10–12		3
1.2–1.4	1					12–14		2
						14–16	1	2
Totals	48	50	Totals	48	50	Totals	48	50

*Each interval, except 0 (exactly), excludes its upper endpoint. Thus a 2,000-word paper in which *commonly* appears twice gives rise to a rate of 1.0 per 1,000 exactly, and the paper appears in the count for the 1.0–1.2 interval.

†With the given lengths of the papers used, it accidentally happens that a rate in this interval cannot occur. For example, if a paper has 2,000 words, a rate of 1 per 1,000 means 2 words, and a single occurrence means a rate of 0.5 per 1,000. Hence a 2,000-word paper cannot lead to a rate per thousand greater than 0 and less than 0.5.

Source: Mosteller and Wallace (1984).

The word *war* has spectacularly different behavior. Although absent from half of Hamilton's papers, when present it is used frequently—in one paper at a rate of 14 per 1,000 words. *The Federalist* papers deal with specific topics in the Constitution and huge variations in the rates of such words as *war, law, executive, liberty,* and *trade* can be expected according to the context of the paper. Even though Madison uses *war* considerably more often than Hamilton in the undisputed papers, we explain this more by the division of tasks than by Madison's predilection for using *war.* Data from use of a word like *war* would give the same troublesome sort of evidence that historians have disagreed about over the last hundred years. Indeed, the dispute has continued because evidence from subject and content has been hopelessly inconclusive.

USE OF NONCONTEXTUAL WORDS

For the statistical arguments to be valid, information from meaningful, contextual words must be largely discarded. Such a study of authorship will not then contribute directly to any understanding of the greatness of the papers, but the evidence of authorship can be both strengthened and made independent of evidence provided by historical analysis.

Avoidance of judgments about meaningfulness or importance is common in classification and identification procedures. When art critics try to authenticate a picture, in addition to the historical record, they consider little things: how fingernails and ears are painted and what kind of paint and canvas were used. Relatively little of the final judgment is based upon the painting's artistic excellence. In the same way, police often identify people by their fingerprints, dental records, and scars, without reference to their personality, occupation, or position in society. For literary identification, we need not necessarily be clever about the appraisal of literary style, although it helps in some problems. To identify an object, we need not appreciate its full value or meaning.

What noncontextual words are good candidates for discriminating between authors? Most attractive are the filler words of the language: prepositions, conjunctions, articles. Many other more meaningful words also seem relatively free from context: abverbs such as *commonly, consequently, particularly,* or even abstract nouns like *vigor* or *innovation.* We want words whose use is unrelated to the topic and may be regarded as reflecting minor or perhaps unconscious preferences of the author.

Consider what can be done with filler words. Some of these are the most used words in the language: *the, and, of, to,* and so on. No one writes without them, but we may find that their rates of use differ from author to author. Table 2 shows the distribution of rates for three prepositions—*by, from,* and *to.* First, note the variation from paper to paper. Madison uses *by* typically about 12 times per 1,000 words, but sometimes he has rates as high as 18 or as low as 6. Even on inspection though, the variation does not obscure Madison's systematic tendency to use *by* more often than Hamilton does. Thus low rates for *by* suggest Hamilton's authorship, and high rates Madison's. Rates for *to* run in the opposite direction. Very high rates for *from* point to Madison but low

Table 2 Frequency distribution of rate per 1,000 words in 48 Hamilton and 50 Madison papers for *by, from,* and *to*

By			*From*			*To*		
Rate per 1,000 Words	*H*	*M*	*Rate per 1,000 Words*	*H*	*M*	*Rate per 1,000 Words*	*H*	*M*
1–3*	2		1–3*	3	3	20–25*		3
3–5	7		3–5	15	19	25–30	2	5
5–7	12	5	5–7	21	17	30–35	6	19
7–9	18	7	7–9	9	6	35–40	14	12
9–11	4	8	9–11		1	40–45	15	9
11–13	5	16	11–13		3	45–50	8	2
13–15		6	13–15		1	50–55	2	
15–17		5				55–60	1	
17–19		3						
Totals	48	50	Totals	48	50	Totals	48	50

*Each interval excludes its upper endpoint. Thus a paper with a rate of exactly 3 per 1,000 words would appear in the count for the 3–5 interval.

Source: Mosteller and Wallace (1984).

rates give practically no information. The more widely the distributions are separated, the stronger the discriminating power of the word. Here, *by* discriminates better than *to,* which in turn is better than *from.*

PROBABILITY MODELS

To apply any of the theory of statistical inference to evidence from word rates, we must construct an acceptable probability model to represent the variability in word rate from paper to paper. Setting up a complete model for the occurrence of even a single word would be a hopeless task, for the fine structure within a sentence is determined in large measure by nonrandom elements of grammar, meaning, and style. But if our interest is restricted to the rates of use of one or more words in blocks of text of at least 100 or 200 words, we expect that detailed structure of phrases and sentences ought not to be very important. The simplest model can be described in the language of balls in an urn, so common in classical probability. To represent Madison's usage of the word *by,* we suppose there is a typical Madison rate, which would be somewhere near 12 per 1,000, and we imagine an urn filled with many thousands of red and black balls, with the red occurring in the proportion 12 per 1,000. Our probability model for the occurrence of *by* is the same as the probability model for successive draws from the urn, with a red ball corresponding to *by* and a black ball corresponding to all other words. To extend the model to the simultaneous study of two or more words, we would need balls of three or more colors. No grammatical structure or meaning is a part of this model, and it is not intended to represent behavior within sentences. What is desired is that it explain the variation in rates—in counts of occurrences in long blocks of words, corresponding to the essays.

We tested the model by comparing its predictions with actual counts of word frequencies in the papers. We found that while this urn scheme reproduced variability well for many words, for other words additional variability was required. The random variation of the urn scheme represented most of the variation in counts from one essay to another, but in some essays the authors changed their basic rates a bit. We had to complicate the theoretical model to allow for this, and the model we used is called the ***negative binomial distribution.***

The test showed also that pronouns like ***his*** and ***her*** are exceedingly unreliable authorship indicators, worse even than words like *war.*

INFERENCE AND RESULTS

Each possible route from construction of models to quantitative assessment of, say, Madison's authorship of some disputed paper, required solutions of serious theoretical statistical problems, and new mathematics had to be developed. A chief motivation for us was to use the *Federalist* problem as a case study for comparing several different statistical approaches, with special attention to one, called the Bayesian method, that expresses its final results in terms of probabilities, or odds, of authorship.

By whatever methods are used, the results are the same: overwhelming evidence for Madison's authorship of the disputed papers. For only one paper is the evidence more modest, and even there the most thorough study leads to odds of 80 to 1 in favor of Madison.

Figures 1 and 2 illustrate how the 12 disputed papers fit the distributions of Hamilton's and Madison's rates for two of the words finally chosen as discriminators. In Figure 1 the top two histograms portray the data for *by* that was given earlier in Table 2. Madison's rate runs higher on the average. Compare the bottom histogram for the disputed papers first with the top histogram for Hamilton papers, then with the second one for Madison papers. The rates in the disputed papers are, taken as a whole, very Madisonian, although 3 of the 12 papers by themselves are slightly on the Hamilton side of the typical rates. Figure 2 shows the corresponding facts for *to.* Here again the disputed papers are consistent with Madison's distribution, but further away from the Hamilton behavior than are the known Madison papers.

Table 3 shows the 30 words used in the final inference, along with the estimated mean rates per thousand in Hamilton's and Madison's writings. The groups are based upon the degree of contextuality anticipated by Mosteller and Wallace (1984) prior to the analysis.

The combined evidence from nine common filler words shown as group B was huge—much more important than the combined evidence from 20 low-frequency marker words like *while-whilst.* These 20 are shown as groups C, D, and E.

There remains one word that showed up early as a powerful discriminator, sufficient almost by itself. When should one write *upon* instead of *on*? Even authoritative books on English usage don't provide good rules. Hamilton and

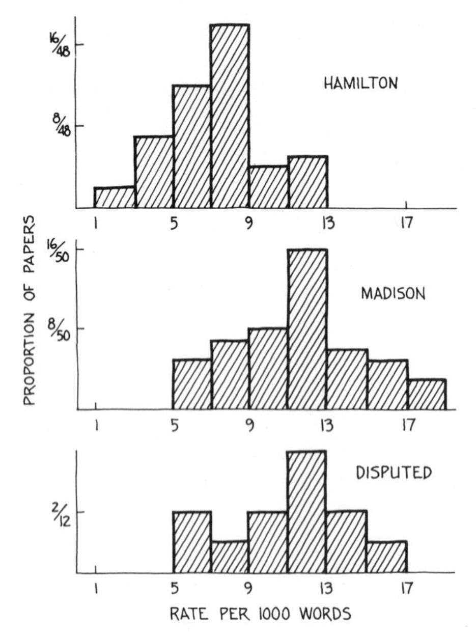

Figure 1 *Distribution of rates of occurrence of* by *in 48 Hamilton papers, 50 Madison papers, 12 disputed papers.*

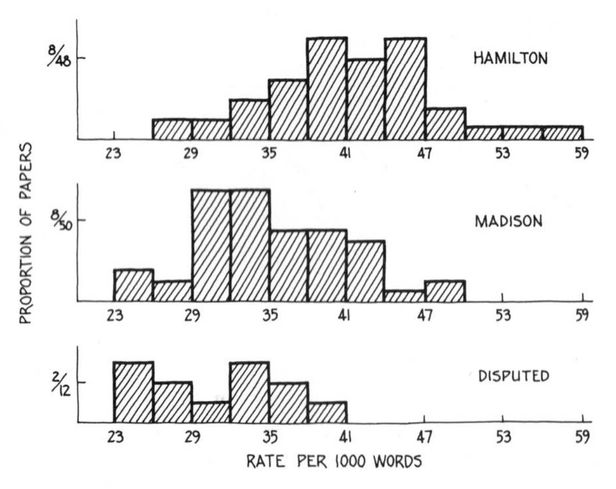

Figure 2 *Distribution of rates of occurrence of* to *in 48 Hamilton papers, 50 Madison papers, 12 disputed papers.*

Madison differ tremendously. Hamilton writes *on* and *upon* almost equally, about 3 times per 1,000 words. Madison, on the other hand, rarely uses *upon*. Table 4 shows the distributions for *upon*. In 48 papers Hamilton never failed to use *upon;* indeed, he never used it less than twice. Madison used it in only 9 of 50 papers, and then only with low rates. The disputed papers are clearly Madisonian with *upon* occurring in only 1 paper. That paper, fortunately, is strongly classified by the other words. It is not the paper with modest overall odds.

Table 3 Words used in final discrimination and adjusted rates of use in text by Madison and Hamilton

Word	*Rate Per 1,000 Words*		*Word*	*Rate Per 1,000 Words*	
	Hamilton	*Madison*		*Hamilton*	*Madison*
Group A			*Group D*		
Upon	3.24	0.23	Commonly	0.17	0.05
Group B			Consequently	0.10	0.42
Also	0.32	0.67	Considerable(ly)	0.37	0.17
An	5.95	4.58	According	0.17	0.54
By	7.32	11.43	Apt	0.27	0.08
Of	64.51	57.89	*Group E*		
On	3.38	7.75	Direction	0.17	0.08
There	3.20	1.33	Innovation(s)	0.06	0.15
This	7.77	6.00	Language	0.08	0.18
To	40.79	35.21	Vigor(ous)	0.18	0.08
Group C			Kind	0.69	0.17
Although	0.06	0.17	Matter(s)	0.36	0.09
Both	0.52	1.04	Particularly	0.15	0.37
Enough	0.25	0.10	Probability	0.27	0.09
While	0.21	0.07	Work(s)	0.13	0.27
Whilst	0.08	0.42			
Always	0.58	0.20			
Though	0.91	0.51			

Source: Mosteller and Wallace (1984).

Table 4 Frequency distribution of rate per 1,000 words in 48 Hamilton, 50 Madison, and 12 disputed papers for *upon*

Rate per 1,000 Words	*Hamilton*	*Madison*	*Disputed*
0 (exactly)*		41	11
0⁺–0.4		2	
0.4–0.8		4	
0.8–1.2	2	1	1
1.2–1.6	3	2	
1.6–2.0	6		
2.0–3.0	11		
3.0–4.0	11		
4.0–5.0	10		
5.0–6.0	3		
6.0–7.0	1		
7.0–8.0	1		
Totals	48	50	12

*Each interval, except 0 (exactly), excludes its upper endpoint. Thus a paper with a rate of exactly 3 per 1,000 words would appear in the count for the 3.0–4.0 interval.

Source: Mosteller and Wallace (1984).

Of course, combining and assessing the total evidence is a large statistical and computational task. High-speed computers were employed for many hours in making the calculations, both mathematical calculations for the theory and empirical ones for the data.

You may have wondered about John Jay. Might he not have had a hand in the disputed papers? Table 5 shows the rates per thousand for nine words of highest frequency in the English language measured in the writings of Hamilton, Madison, Jay, and, for a change of pace, in James Joyce's *Ulysses.* The table supports the repeated assertion that Madison and Hamilton are similar. Joyce is much different, but so is John Jay. The words *of* and *to* with rate comparisons 65/58 and 41/35 were among the final discriminators between Hamilton and Madison. See how much more easily Jay could be discriminated from either Hamilton or Madison by using *the, of, and, a,* and *that.* The disputed papers are not at all consistent with Jay's rates, and there is no reason to question his omission from the dispute.

SUMMARY OF RESULTS

Our data independently supplement the evidence of the historians. Madison is extremely likely, in the sense of degree of belief, to have written the disputed *Federalist* papers, with the possible exception of paper 55, and there our evidence yields odds of 80 to 1 for Madison—strong, but not overwhelming. Paper 56, next weakest, is a very strong 800 to 1 for Madison. The data are overwhelming for all the rest, including the two papers historians feel weakest about, papers 62 and 63.

For a more extensive discussion of this problem, including historical details, discussion of actual techniques, and a variety of alternative analyses, as well as for a brief review of authorship studies since 1969, see Mosteller and Wallace (1984).

Table 5 Word rates for high-frequency words (rates per 1,000 words)

	*Hamilton (94,000)**	*Madison (114,000)**	*Jay (5,000)**	*Joyce (Ulysses) (260,000)**
The	91	94	67	57
Of	65	58	44	30
To	41	35	36	18
And	25	28	45	28
In	24	23	21	19
A	23	20	14	25
Be	20	16	19	3
That	15	14	20	12
It	14	13	17	9

*The number of words of text counted to determine rates.

Sources: Hanley (1937); Mosteller and Wallace (1984).

PROBLEMS

1. Why can't the authorship of the disputed papers be determined by literary style or political philosophy?
2. a. What is a discriminator?
   b. Distinguish at least two categories of discriminators.
   c. Why is *by* a good discriminator? (Refer to Table 2.)
3. What is a "noncontextual word"?
4. Why do the authors use word frequency per thousand words instead of just the number of occurrences?
5. Refer to Table 1. In how many of the Hamilton papers studied does the word *commonly* appear at least once?
6. Refer to Table 2. In what percentage of the Madison papers studied does *from* occur 3–7 times per 1,000 words? (Note: The interval 3–7 uses the authors' convention on intervals.)
7. Consider the "balls in an urn" model. How many colors of balls would we need to extend the model to the simultaneous study of five words? Of *n* words?
8. Consider Figure 1. True or false: More than ⅓ of the Hamilton papers studied use *by* 3–7 times per 1,000 words.
9. Study Figure 2. Does the graph for the disputed papers look more like the graph for the Hamilton or the Madison papers?
10. Consider Table 3. Looking at group B, which word would you say was the best Hamilton/Madison discriminator? What was your word-selection criterion? Answer the same questions for group D.
11. Table 1 shows the relative frequency of *war*. Why doesn't *war* appear in Table 3?

REFERENCES

Miles L. Hanley. 1937. *Word Index to James Joyce's "Ulysses."* Madison, Wis.: University of Wisconsin.

F. Mosteller and D. L. Wallace. 1964. *Inference and Disputed Authorship: The Federalist.* Reading, Mass.: Addison-Wesley.

F. Mosteller and D. L. Wallace. 1984. *Applied Bayesian and Classical Inference: The Case of "The Federalist Papers"* (2nd ed. of *Inference and Disputed Authorship: The Federalist).* New York: Springer-Verlag.

Children's Recall of Pictorial Information

Doris R. Entwisle *Johns Hopkins University*
W. H. Huggins *Johns Hopkins University*

Finding out how children mature and develop, and what helps or hinders their growth, is important for pediatricians, teachers, parents, and, above all, for the children themselves. Although we know more about how children's capabilities develop than we used to, we still do not know very much. When children leave the sixth grade, for example, their basic reading and math skills commonly range from first-grade to tenth-grade competence, a span of more than nine years. Why some children develop so much better than others is unclear. Perhaps children have more competence along some lines than they are credited with, and perhaps children get assistance from some kinds of human talents that so far we have not recognized. Clearly, in assessing children's competence we take a narrow view if, as is usually true, we concentrate only on the kinds of abilities that are tapped by standardized tests of reading and mathematics. These tests tell us where children are with respect to these skills but not how they got there.

Two articles that came to our attention suggested that children might have some capabilities of a visual nature that had been overlooked, particularly in their ability to recall pictorial information. One was Haber's (1970) research, showing that after a few seconds' exposure to more than 2,500 slides, adults later identified correctly 85% to 95% of the "old" slides that were paired with slides not previously seen. The other report (Kagan, 1970) provided a hint of the early development of the ability that Haber had discovered in adults, in that a 4-year-old correctly recognized a large proportion of pictures taken from magazines she had previously seen. Also, of course, young children quickly learn to use visual information to find their way around the neighborhood or other people's houses.

To start with, we did a pilot study, and this is where statistics first comes in. We showed photographic slides in rapid succession to 10 nursery school children aged 3 and 4. The slides were of landscapes and other scenes unfamiliar to the children and did not show human beings. Each child viewed 5 slides successively in a hand-viewer. Immediately afterward 10 slides were presented sequentially, with the 5 viewed initially randomly intermixed with 5 new ones. Some of these children could correctly identify as "new" or "old" 9 of the 10 slides. The question is whether this is unusual—whether 3- and 4-year-olds did have some ability or whether this number of correct choices could have easily occurred by chance.

To answer this question we used a binomial model. Getting 9 out of 10 slides correct is just like getting 9 out of 10 heads in a series of coin tosses. Table 1 shows how the binomial model works for 10 outcomes. We find that 9 out of 10 *is* unusual—it could occur only 11 out of 1,024 times. (We would be impressed also if we observed 10 out of 10 heads, so we add up the chances of 9 or more choices being correct.) Even children as young as could be tested for picture recognition were quite good at it. We then went on to set up a test to use with first-graders that drew upon the same general strategy.

Table 1 Likelihood of various outcomes for 10 coin tosses

Outcome	*Chances out of 1,024*	*Percentage of Time Outcome is Observed*
0 heads, 10 tails	1	0.1
1 head, 9 tails	10	1.0
2 heads, 8 tails	45	4.4
3 heads, 7 tails	120	11.7
4 heads, 6 tails	210	20.5
5 heads, 5 tails	252	24.6
6 heads, 4 tails	210	20.5
7 heads, 3 tails	120	11.7
8 heads, 2 tails	45	4.4
9 heads, 1 tail	10	1.0
10 heads, 0 tails	1	0.1
Totals	1,024	100.0

Table 2 Likelihood of various outcomes for 40 coin tosses

Outcome	*Percentage of Time Outcome is Observed*
40 heads, 0 tails to 31 heads, 9 tails	0.03
30 heads, 10 tails	0.1
29 heads, 11 tails	0.2
28 heads, 12 tails	0.5
27 heads, 13 tails	1.1
26 heads, 14 tails	2.1
25 heads, 15 tails	3.7
24 heads, 16 tails	5.7
23 heads, 17 tails	8.1

The second use of statistics is in how we designed the study of first-graders. We wanted to find out whether children could recall slides they had seen previously—again, yes or no, so the binomial model fits. Since they could guess and be right half of the time, we had to present them with a large enough number of slides to get a good test. For example, children could have some ability and get, say, 7 out of 10 correct, but, as Table 1 shows, 7 out of 10 can happen often enough to be judged a fairly common event. If we showed 40 slides, though, the "chance" (50/50) baseline is 20, and there is plenty of room between 20 and 40 for children to display their ability. To get 40 out of 40 would happen only once in 2^{40} times, analogously with getting 10 out of 10, which Table 1 shows would happen only once in 2^{10} times (1,024). But 2^{40} is $(1{,}024)^4$, so to get 40 out of 40 would happen only once in more than 1,000,000,000,000 times. To evaluate the possible outcomes with 40 coin tosses would be tedious but fortunately tables of the binomial are available for various numbers of trials.

Table 2 shows the relevant portion of a binomial table calculated for 40 outcomes. The occurrence of any outcome of 31 or more heads will be rare—altogether these outcomes will happen only 0.03% of the time—so we have abbreviated the table. If we decide on a chance of "5% or less" as our criterion for performance rare enough to be "unlikely," we can pick the point in Table 2 where something close to a 5% cutoff occurs. The chances of 26 or more heads add to a little more than 4% (0.03 + 0.1 +0.2 + 0.5 + 1.1 + 2.1). We see also that 25 heads could occur 3.7% of the time. If we chose the cutoff at 25, the chance probability would be about 4.0 + 3.7 or 7.7%, which is greater than our acceptable risk of 5%. We therefore settled on 26 heads or more as the number closest to our "ideal."

THREE EXPERIMENTS

With statistics thus used to figure out a good design, we then went ahead and did three experiments that all followed this design but each was pointed at a slightly different question.

1. *Condition I: Black-and-White Slides.* A class of 23 first-grade children of slightly higher than average intelligence (IQ = 109) was shown 40 black-and-white slides. The children were instructed to "look at these pictures we are going to show you because we may ask you some questions about them later on." The slides were prints of unremarkable landscapes or cityscapes and were completely unfamiliar to these children, insofar as we knew.

Each slide was displayed for 10 seconds, with 3 seconds intervening between slides. We started at 10 A.M. and continued until all 40 slides had been shown. At 12:30 P.M. on the same day two screens were set up side by side in the front of the classroom, and two projectors were used to display 40 pairs of slides for 10 seconds per pair, with 3 seconds intervening. Each of the 40 slides shown earlier was paired randomly on the left or right with 40 new slides of the same type. The children were instructed to indicate in special booklets which slide of each pair they had seen earlier. As Table 3 shows, the children correctly identified on the average 31 of the slides. So children do have some ability because 26 or more correct identifications for a single child is significant with a binomial model, as we figured out in advance.

2. *Condition II: Colored Slides.* Condition II was identical with Condition I except that a different class of 30 first-grade children of slightly higher than average intelligence (IQ = 109) participated, and the slides were *colored.* (The slides used in Conditions I and II were identical except for the presence of color in Condition II.) The children on the average correctly identified more than 34 of the colored slides (Table 3).

3. *Condition III.* A third class of 29 first-grade children, again of slightly higher than average intelligence (IQ = 109), was shown the same 40 colored slides used as the target slides in Condition II. Condition III was the same as Condition II except that a week was allowed to elapse before testing recognition. On the average, the children identified more than 31 of these slides correctly.

Table 3 Slide identification

	Condition I (Black-and-White Slides, 2½-Hour Delayed Test, 23 children)	*Condition II (Colored Slides, 2½-Hour Delayed Test, 30 children)*	*Condition III (Colored Slides, 1 Week Delayed Test, 29 children)*
Average number of slides identified correctly	31.0	34.2	31.9
Range of gains above 20	3–18	5–20	4–19
Number and % of individual children scoring above chance level of 20	21 (91%)	29 (97%)	25 (86%)

Note first that in every condition the average number correct is far beyond even the 5% risk we were willing to tolerate. But "average performance" for a group might mean, for example, for Condition I that, of the 23 children, some did very well—say, 40 out of 40—while others scored at chance level. Thus if 13 children got 40 out of 40 correct, and 10 children got 20 out of 40 correct, we would still observe an excess of 11 slides *on average.* Perhaps half the population have this ability and half do not. The range of individual scores, though, as shown in Table 3, shows that almost every child scored above chance level under every condition.

Condition II (colored slides, short delay) looks best. To see if this superiority is "real" we resorted to some complicated statistical procedures that enlightened us as to whether the superiority of condition II over conditions I and III is too strong for chance fluctuations to explain. The answer was yes—this difference favoring condition II is large enough to signify a "real" difference when all the data are studied jointly.

This experiment with first-graders, then, agrees with our earlier informal study of nursery school children. Young children can apparently process and retain a rather large amount of complex pictorial information unrelated in any way to their ongoing activities; this finding is consistent with both Kagan's and Haber's observations. The conditions under which the present experiment was carried out probably led to underestimates of children's absolute ability at the recognition task because some children were not paying close attention all the time. If they had been concentrating—for example, if we had tested them individually in a room with no distractions—they would probably have done better, but the conditions of these experiments resemble those in ordinary classrooms where similar activities occur in teaching.

FURTHER WORK

Other, later work (Feinman and Entwisle, 1976) suggests that pictorial information processing ability may be as good in young children as in adults. When some third-graders were asked to view and recognize slides depicting human faces with the same general approach as that described above for the experiments showing landscapes, their performance was about the same as that of college students.

How children store pictorial information in memory is an important matter for future investigation. Clearly, color enhances performance in that it makes up for a week's forgetting time, but this may happen because it gets children's attention better rather than because it changes how information is stored.

One explanation for why children can recognize so many of the slides correctly lies in the richness of the information present in the slide. A slide contains information that can be remembered in terms of many different features. A harbor scene, for instance, has water, a shoreline with a particular configuration, boats in a spatial arrangement, and so on. If it is the only harbor scene in the set of slides, then coding it as a "harbor scene" is sufficient information to allow it to be distinguished from all other slides in the set. But if the whole set of slides consists entirely of harbor scenes, then the task is enormously more difficult. We could explain the effects observed in our experiments by noting that the slides were different enough from one another so that remembering one or two features in any one slide would be sufficient to recognize it. For example, harbor scenes were different enough to be quite distinctive.

The point of this paper is to call attention to the rather remarkable ability of young children to recognize slides after a considerable time and to emphasize that pictures are potentially capable of conveying a great amount of information that children retain. The retention over a one-week period is especially surprising. This could be an important factor in the education of preliterate children, or for children who for various reasons, including dialect problems, do not develop reading facility as well or as early as others.

PROBLEMS

1. How could the materials used in these experiments be tied more closely to the school curriculum?
2. Do you think first-grade children would be better or worse at recalling human faces and figures than recalling landscapes?
3. How long do you think significant memory effects would persist past one week? Explain your answer.

REFERENCES

S. Feinman and D. R. Entwisle. 1976. "Children's Ability to Recognize Other Children's Faces." *Child Development* 47: 506–510.

R. N. Haber. 1970. "How We Remember What We See." *Scientific American* 220: 104–112.

J. Kagan. 1970. "The Determinants of Attention in the Infant." *American Scientist* 58: 298–306.

The Meaning of Words

Joseph B. Kruskal *Bell Telephone Laboratories, Inc.*

How can we possibly use statistics to capture that elusive thing called "meaning"? Even granting the possibility, won't it take all the romance out of poetry and the charm out of graceful speech?

Well, we *can* study the meaning of words by the orderly methods of statistics. I shall discuss two interesting studies of this kind. But the romance of words, you will discover, is safe from science.

Practical-minded people may ask *why* we should work so hard to pin down meaning. Doesn't language, after thousands of years of natural evolution, serve well enough? Quite simply, no. Any college admissions officer will testify to the difficulty of interpreting teachers' written recommendations, and many colleges use a standard form to reduce this difficulty. Studies like those discussed here may one day improve communication among teachers, students, and colleges. They may even contribute to peace among nations.

Numbers, in other situations, have supplemented descriptive words for a long time, often very helpfully. The use of ***inches, pounds,*** and ***degrees Fahrenheit*** certainly has advantages over ***fairly short, very light,*** and ***rather hot.*** Once upon a time, ***feverish*** carried a meaning as vague as ***charming*** or ***surly.*** The introduction of the fever thermometer has been a boon to human health.

In subtler areas, such as musical tones, speech sounds, and intelligence, numbers have wide use today. Color provides a striking example. Many people imagine colors so subtle as to elude numerical description, yet several numerical systems enjoy routine use. (In most of them, three numbers describe each color.) (See the essay by Vance on determination of numerical color tolerances.) These systems played a vital role in developing color television and improving color photography, and have also aided the paint industry, stage makeup, and other fields. Some artists, too, find this scientific way of describing colors helpful. Nothing about this science reduces the scope of their art, nor your pleasure in looking at their paintings, seeing a color movie, or watching the sun go down.

These older examples, however, differ from the studies to be described here in one important way: the manner in which measurements are taken. Physical devices can measure height, weight, and temperature. Colors are handled by elaborate devices for mixing and dimming light, and a standard test measures intelligence. In the work described below, however, a person acts as a measuring instrument rather than as an object of study. His or her subjective impression, obtained and analyzed in a very careful way, constitutes the measurement.

PERSONALITY TRAITS

A study by Rosenberg, Nelson, and Vivekananthan (1968) deals with the meaning of 64 personality-trait words such as ***impulsive, sincere, cautious, irritable,*** and ***happy***—words that describe people. By using a novel statistical technique the study helps to provide order where order is difficult to find.

Everyone uses these words, and they carry meaning to us all. Furthermore, they have importance. Imagine overhearing someone describe you as "humorless" or as "good-natured," and think how different these two descriptions would make you feel! When these words are used to arrange blind dates, or by teachers writing recommendations, it makes a big difference just which ones are chosen.

Nevertheless, people use these words differently. If two people both describe the same friend, or the same movie character, we would be surprised if they used *exactly* the same words. In psychiatry, the same uncertainty of meaning exists, not only for common words but also for technical terms such as ***schizoid*** and ***autistic.*** Long articles are written about what these words should or do mean.

This creates the problem. With words that are used so frequently and carry so much importance, clarification of their meaning may well prove useful—to psychiatrists perhaps, to college admissions officers possibly, to computer dating organizations, and who knows to whom else.

A DETOUR ON MULTIDIMENSIONAL SCALING

Clarify their meaning—but how? Statistics first enters here, providing an approach to the problem, as well as a method of analysis. First, however, let us take stock of the problem more clearly.

1. What do we *want*? An orderly description of personality-trait words, according to their meaning. With success, we will obtain a description as helpful as a city map, which displays roads, bus routes, parks, and so forth, according to their location.
2. What *means* do we have for getting this description? Nothing but the way people use the words. We have nothing to measure with a ruler or a thermometer; we can only observe how people use or respond to the words.

A recently developed statistical method called *multidimensional scaling* has particular value in problems like this one. The fact that Rosenberg and his coauthors were aware of this method helped stimulate their experiment. Statistics provided the approach.

What does multidimensional scaling consist of? It can best be described in three parts: what goes in, what comes out, and what happens in between. What goes in to multidimensional scaling are *similarities* or *dissimilarities* between various objects of one kind. For each pair of objects (in this case, for each pair of words from among the 64 used), a number describes how much alike, or how different, the two objects are. Many methods can provide such numbers. For example, the similarity between two colors may be provided by a person who looks at the two colors and describes, on a scale from 1 to 5, how alike they are. (If a larger number indicates greater likeness we call the numbers "similarities"; in the opposite case, "dissimilarities.") A more elaborate method was used in this study; I'll describe it later.

What comes out of multidimensional scaling is a *map* or *picture,* such as Figure 1. (For the moment ignore the lines in the figure and consider only the configuration of the points.) Briefly, the method places each object in a particular position. The map produced by scaling, though not a real map, shares many characteristics of real ones. For example, we are free to turn it and look at it from any direction we please; it has no particular direction that is truly up. Also we are free to enlarge (or diminish) the map, change the scale, so to speak, though our maps do not really have any scale. Convenience dictates how big we make the map.

What happens between the input and the output of multidimensional scaling? It is simple to say, though hard to do. We place objects on a map with the goal of having objects that are close together be very much alike and having objects that are far apart be very different. In other words, small distances should correspond to small dissimilarities, and large distances to large dissimilarities; or vice versa for similarities. In brief, the goal while constructing the map is a good relationship between the map distances and the input (dissimilarities or similarities).

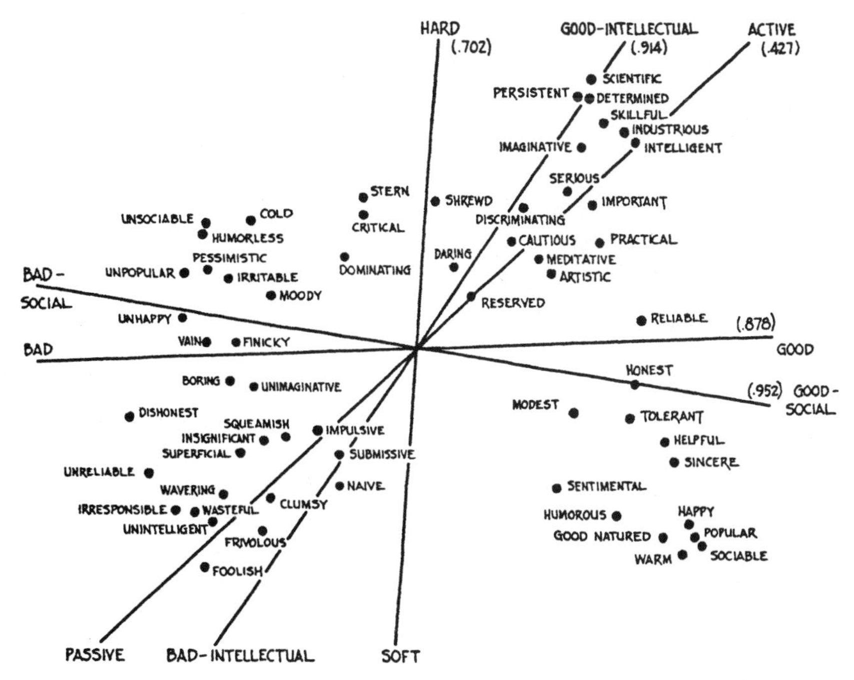

Figure 1 *Multidimensional scaling map with trend axes. Source: Rosenberg, Nelson, and Vivekananthan (1968)*

How good a relationship can be achieved depends on the input. The quality of the relationship has great importance in multidimensional scaling. Where it is good, the map may tell us something useful; where it is very bad, we may as well throw the map away.

The complex process of constructing the map is almost always carried out by computer. The user of the computer program does not need to understand this process any more than the driver of a car needs to understand how its motor works. (Today multidimensional scaling is frequently carried out on ordinary personal computers.)

PERSONALITY TRAITS RESUMED

To obtain the similarities, Rosenberg and his coauthors presented 69 subjects (college students) with slips of paper containing 64 personality-trait words. Each slip contained one word, and each subject received a complete pack of 64 slips. The subjects were asked to put the slips in their packs into roughly 10 groups, so that the words in a group could plausibly describe a single person. (A slip could be put in only one group.) Of several procedures that psychologists use to obtain similarities, this is the most rapid.

By counting how many subjects put two words together, the investigators got an *agreement score* for the two words. For example, 37 of the subjects put *reliable* and *honest* in the same group, so these two words have an agreement score of 37. Several examples are:

1. ***Reliable*** and ***honest,*** score = 37
2. *Clumsy* and *naive,* score = 19
3. *Sentimental* and ***finicky,*** score = 2
4. *Good-natured* and ***irritable,*** score = 1
5. *Popular* and *unsociable,* score = 0

The agreement scores form a 64-by-64 square array. The 64 agreement scores of a word with itself are of no interest because all equal 69, the number of subjects. The remaining $64 \times 64 - 64 = 4{,}032$ values come in pairs of equal values because the agreement score of *warm* with ***intelligent,*** for example, is just the same as the agreement score of ***intelligent*** with *warm.* Thus only $4{,}032/2 = 2{,}016$ distinct agreement scores really matter.

The agreement scores are similarities and can be used as the input for multidimensional scaling. In fact, the authors tried this in the earlier stages of their analysis, but the results were not clear. Fortunately the authors spotted the trouble: many of the agreement scores are so small that comparisons among them have little reliability. The scores given above shout that pair 2 is more alike than pair 3, but they only whisper that pair 3 is more alike than pair 4. With more than half the scores under 10, greater accuracy could help a lot. (Incidentally, this sort of trial and improvement often occurs when data are analyzed in a novel way.)

Fortunately, greater accuracy was found by using the data differently. For example, consider any two words; say, *sentimental* and ***finicky.*** Let's now use two rows to list their agreement scores with all 64 words:

	Foolish	*Inventive*	*Wavering*	*Submissive*	*Cold*	*Tolerant . . .*
Sentimental	2	6	5	15	3	22 . . .
Finicky	6	2	18	10	9	2 . . .

If two words have similar meanings, then not only should they have a large agreement score, but also ***their agreement scores with a third word should be about the same.*** Thus for two similar words the corresponding numbers in the two rows should not differ greatly. If we measure the difference between these two rows of numbers in some reasonable way, then this secondary (or derived) measure should indicate the dissimilarity (rather than similarity) in meaning of the two words. Following this idea, the dissimilarity between *sentimental* and ***finicky*** is formed by adding up the squares of the differences,

$$(2 - 6)^2 + (6 - 2)^2 + (5 - 18)^2 + (15 - 10)^2 + (3 - 9)^2 + (22 - 2)^2 + \dots .$$

The larger this dissimilarity, the greater the difference between the words.

The application of multidimensional scaling to the dissimilarities[1] gave the map in Figure 1. This map must be examined, interpreted, and assessed before we can feel we have found something useful or illuminating. You will find it profitable to examine Figure 1, to see what groups of words occur together, and to see how meanings change systematically across the figure. If an investigator finds anything sensible at this point, then a preliminary judgment of success may be entered. If the map provides an orderly picture that we did not know about before, that is progress.

PERSONALITY TRAITS—A SECOND STEP

The authors found a good deal of structure in Figure 1:

> Going from the upper-right corner to the lower-left corner, the desirability of traits for intellectual activities appears to decrease systematically. Another systematic change appears to take place as one goes from the upper-left corner to the lower-right corner; in this case the social desirability of the traits increases.

Notice that the upper-right corner includes ***intelligent, skillful,*** and ***scientific,*** while the opposite corner includes ***unintelligent, foolish,*** and ***frivolous.*** The upper-left corner includes ***unsociable, humorless,*** and ***unpopular,*** while the opposite corner includes ***sociable, popular,*** and ***warm.*** A good way to indicate these trends is by axes, as drawn in Figure 1. For each trait, its position *along* the axis indicates where it stands with regard to that trend. (To get the position of a word along the axis we drop a perpendicular onto the axis from the plotted position of the word. It makes no difference how far from the axis the word lies.)

This interpretation does not exclude other interpretations. A given direction (such as upper right to lower left) may have several interpretations, and other directions may have meaning also. Not everybody agrees on the best way to interpret the figure.

The authors did not stop with their own subjective interpretation. Subjects different from those who provided the agreement scores rated the 64 traits on five different scales. Each subject dealt only with a single scale. For example, 34 subjects rated each of the traits on a 7-point scale "according to whether a person who exhibited each of the traits would be good or bad in his intellectual activities." A second scale dealt with social activities. Three other scales, namely, good-bad, hard-soft, and active-passive, were also used. (These were chosen from among the semantic differential scales in Osgood, Suci, and Tannenbaum, 1967.)

For each scale, the authors took several steps. They found the median of the subject ratings for each trait. (The median of the ratings is the middle value,

[1]Due to limitations (no longer present) on the computer program, only 60 words were actually included in the multidimensional scaling.

smaller than half the values and larger than half the values.) By using a statistical method called *linear regression,* they found the axis that best matches the median ratings. The five axes shown in Figure 1 resulted from this procedure. The second experiment clearly verifies the two trends described by the authors, as well as displaying some other trends. (There may still be other valid trends.)

It is instructive to check the strength of each trend. In other words, how well do the median ratings on each scale match the positions of the words along the corresponding axis? To make this comparison, we calculate the *correlation coefficient* between the median scores for the word on the scale, and their positions on the axis. This widely used statistical measurement indicates how closely two sets of numbers vary with each other. It always lies between −1 and +1, and +1 indicates perfect agreement in the way the numbers vary. The correlation coefficients, shown in Figure 1, are:

Social good-bad	.95
Intellectual good-bad	.91
Good-bad	.88
Hard-soft	.70
Active-passive	.43

These indicate that the first three scales match the map really quite well, the fourth only fairly well, and the last in a very mediocre way (although .43 is high enough to prove conclusively that there is *some* connection between this scale and the map).

In conclusion, this analysis makes a significant start toward providing a systematic explanation of the meanings of a considerable set of words for personality traits. The authors have constructed a single map that explains how subjects use and understand these words in several different tasks. Of course, the map only explains *part* of the meaning of these words. Other aspects are entirely ignored. Thus *skillful, industrious,* and *intelligent* are very close on the map because they often describe the same person, even though their meanings differ greatly in other respects. Also, the map only explains *aggregate* data, based on many subjects. This washes out and ignores individual variation in the use of these words.

Nevertheless, it is striking and illuminating that a simple map captures an important part of the meaning of a wide variety of words for personality traits. Real progress has been made.

MULTIDIMENSIONAL SCALING—A FURTHER COMMENT

The maps produced by multidimensional scaling need not be ordinary two-dimensional maps. They may be (and often are) three-dimensional, four-dimensional, or even higher-dimensional. As a matter of fact, Rosenberg and his coauthors actually constructed a one-dimensional map, a two-dimensional map (Figure 1), and a three-dimensional map, before deciding which one was the right one. (To explain how the choice was made would take us too far afield.)

I mentioned earlier that the quality of the relationship between map distances and dissimilarities has great importance. If the relationship is very bad, the map might as well be thrown away. A picture of this relationship (for the map of Figure 1) appears as the central plot of Figure 2. This plot contains $(60 \times 59)/2 = 1{,}770$ points, one for each pair of words. For each pair, the dissimilarity gives the vertical coordinate of a point, and the distance between the two words (taken from Figure 1, just as you would measure it with a ruler) gives the horizontal coordinate. Clearly quite a good relationship connects dissimilarity and distance. The other two plots of Figure 2 give the corresponding pictures, based on one-dimensional and three-dimensional maps.

NATIONS

The names of nations, like other words, bring to mind many associations. Two experiments [one by Wish (1970) using 12 nations and another by Wish, Deutsch, and Biener (1970, 1972) using 21 nations] investigated how people perceive nations and their interrelationships. The most important questions addressed to the subjects concerned how *similar* various nations are. By methods somewhat like those above, and using students at Columbia University as subjects, the authors discovered several interesting facts.

Of the many characteristics of nations that influence their perceived similarity, it is not surprising that the two that emerge as most influential (in both experiments) are political alignment (ranging from "aligned with U.S.A." to "aligned with Russia") and economic development. A more complex characteristic that also displayed importance in the second experiment is "geography, race, and culture," under which the nations break down into four groups: European, Spanish, African, and Asian.

The importance of the characteristics to different subjects did yield surprises, however. In the first experiment, the 18 subjects were divided into doves, moderates, and hawks according to which of five recommendations each selected concerning the war in Vietnam. All six hawks attached greater impor-

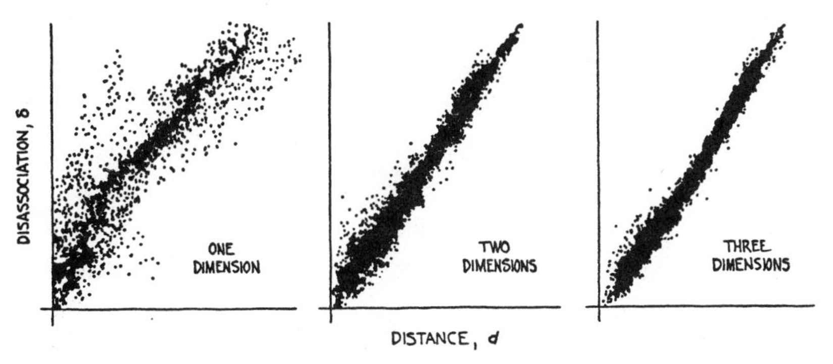

Figure 2 *Relationships between map distances and word dissimilarities. Source: Rosenberg, Nelson, and Vivekananthan (1968)*

tance (in making their similarity judgments) to political alignment, while all seven doves attached greater importance to economic development. Thus political attitude affects even a question like how similar two nations are.

In the second experiment, 75 subjects from eight countries displayed the same systematic effect. Also, females as compared with males and students from underdeveloped countries as compared with students from developed countries attached higher importance to political alignment. The students were also tested for their knowledge about the 21 nations and were then divided into better informed and less informed groups. Within each of these groups, the same effects were displayed, although much more strongly among the less informed group.

PROBLEMS

1. In what way are the measurements used in the personality-traits study different from pounds or inches?

2. What quality was being quantified in preparing the input for multidimensional scaling?

3. Suppose Rosenberg and his coauthors had been studying the meaning of 50 (rather than 64) personality-trait words. How many distinct agreement scores would have then mattered?

4. Consider a Rosenberg-type experiment using only four words with the following agreement scores:

	Faulty	*Friendly*	*Mean*	*Perfect*
Faulty	—	1	4	2
Friendly	1	—	0	4
Mean	4	0	—	1
Perfect	2	4	1	—

   What is the dissimilarity between *friendly* and *perfect*?

5. Refer to Figure 1.
   a. What five words form an extreme good-soft group?
   b. How would you characterize *stern* and *critical*?

6. Consider the locations of *meditative* and *impulsive* in Figure 1.
   a. How do the two words compare on the active/passive axis?
   b. Does the answer in (a) seem counterintuitive? Explain your answer.

7. Were the axes in Figure 1 produced by the multidimensional scaling procedure? Explain your answer.

8. The number .914 (under Good-Intellectual) in Figure 1 is a correlation coefficient. What two measures are being correlated? What can you conclude about these two measures?

9. Examine Figure 2. How is δ measured? What about d?

REFERENCES

Charles E. Osgood, George J. Suci, and Percy Tannenbaum. 1967 (1957). *The Measurement of Meaning.* Urbana, Ill.: University of Illinois.

S. Rosenberg, C. Nelson, and P. S. Vivekananthan. 1968. "A Multidimensional Approach to the Structure of Personality Impressions." *Journal of Personality and Social Psychology* 9:283–294.

Myron Wish. 1970. "Comparisons Among Multidimensional Structures of Nations Based on Different Measures of Subjective Similarity." *General Systems* 15:55–65.

Myron Wish, M. Deutsch, and L. Biener. 1970. 'Differences in Conceptual Structures of Nations: An Exploratory Study." *Journal of Personality and Social Psychology* 16:361–373.

Myron Wish, M. Deutsch, and L. Biener. 1972. "Differences in Perceived Similarity of Nations." In *Multidimensional Scaling: Theory and Applications in the Behavioral Sciences,* A. K. Romney and S. Nerlove, eds. New York: Academic Press.

The Sizes of Things

Herbert A. Simon *Carnegie-Mellon University*

On Figure 1 are drawn four lines. The lowest one, a simple straight line inclined at a 45° angle, serves merely for purposes of comparison in describing the three slightly wavy lines. The three wavy lines—and particularly the two just above the straight line—depict some curious facts about the world. Whether they are significant facts as well as curious facts is a question we examine.

The lower broken line relates, on a logarithmic scale,[1] the 1980 populations of the 20 largest cities in the United States to the ranks by size of the cities,

[1]The common logarithm is probably familiar as a tricky device for multiplying numbers through a process of addition. Another way of looking at the logarithm is that taking the logarithm compresses the scale of numbers so as to create a new scale, one that makes multiplying the old number by 10 equivalent to adding one unit to the new number. For example, the logarithm of 10 is 1, of 100 is 2, of 1,000 is 3, and so on. The logarithm of 2,000 is about 3.300 and that of 20,000 is about 4.300. If a city has a population of 5,000,000, then the logarithm of its population is about 6.70. The compression achieved by a logarithm scale increases as the numbers do.

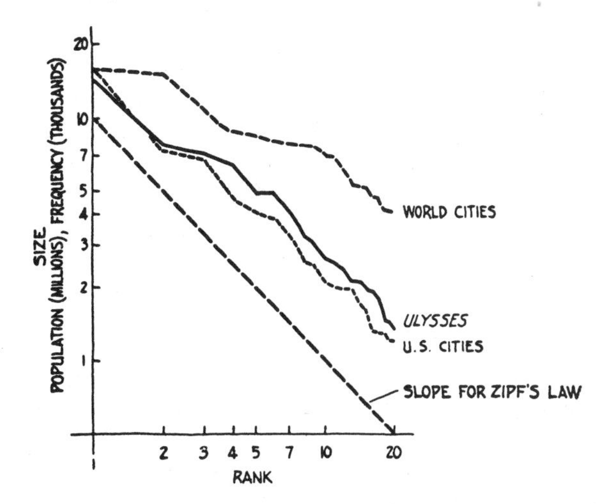

Figure 1 *Logarithm of size plotted against logarithm of rank for frequencies of words and for populations of cities.*

arranged with New York, ranked 1, down to Cincinnati, ranked 20. The population of each metropolitan statistical area was used, not just that within the city limits. The horizontal axis shows, also on a logarithmic scale, the city ranks, from 1 through 20; on the vertical scale are shown the corresponding logarithms of populations in millions of persons. Ignoring the two largest cities (New York and Los Angeles), we can see that the rest of the line is nearly straight and inclined nearly at a 45° angle, parallel to the straight line below. Straightening out the left end of the curve would involve raising New York from about 20 million people to about 30 million and Los Angeles to 15 million (a heavy price to pay for a straight line), but the remaining 18 cities would require very little adjustment—generally less than 10% up or down.

The solid line, just above and very close to the line for cities, shows (again on logarithmic scales) the number of occurrences of each of the 20 words most frequently used in James Joyce's *Ulysses,* when the words are arranged in descending order of frequency of occurrence. For this line, the ordinates show the frequencies of occurrences in thousands. The most frequent word in *Ulysses, the,* occurred 14,887 times; the twentieth most frequent, *all,* occurred 1,311 times. As with the city sizes, the word frequencies lie almost on a straight line, although straightening the line would again require adjustment of the first few words; *the* would have to be increased to about 26,000 occurrences, *of* to 13,000, and *and* to about 8,700. The remaining 17 frequencies are extremely close to a straight line inclined at 45°.

Observe that in these distributions the product of the rank of each item by its size remains constant over the whole scale. If the first item (rank 1) has size 1,000,000, the tenth item will have size about 100,000 (10 × 100,000 =

1,000,000), and the twentieth item will have size 50,000 ($20 \times 50{,}000 = 1{,}000{,}000$). The task before us is to explain why these regularities hold, why the product of number and rank in these distributions is almost constant, and—even more mysterious—why the size distribution of U.S. cities should obey the same law as the frequency distribution of words in a stylistically unusual book such as *Ulysses* (or in any book for that matter). Let's begin with the words.

WORDS: COMMON AND RARE

In the late nineteenth century, several linguists (among them de Saussurre in France) discovered the surprising rank-frequency regularity in the relative contributions of different words to any body of text. Obviously, certain words, such as *of,* will occur rather frequently in almost any English text, while other words, such as *conundrum,* will occur infrequently or not at all. The frequency of any specific word may vary widely from one text to another.

Whenever you arrange the various words occurring in a particular text in the order of their frequency of occurrence—first the word that occurs most often in that text, then the word that occurs next most often, and so on—the regularity depicted in Figure 1 will reappear. The twentieth word on your list will occur about half as often as the tenth word.[2] If you enjoy this kind of numerology, you will find equally startling regularities at the other end of the distribution among the rare words.

About one-half of the total number of *different* words in the text occur exactly once each, about one-sixth occur exactly twice each, and about one-twelfth occur three times each (see Table 1). The ratio $1/[n(n + 1)]$ gives the fraction of all the distinct words in the text that occur exactly n times each. This regularity in frequency of occurrence of the rare words is, of course, the same rank-size law we have been observing at the other end of the distribution, for the rank of a word is simply the cumulated number of different words that have occurred as frequently as it has, or more frequently. Suppose then, as the rank-size rule requires, that $K/(n + 1)$ words occur $n + 1$ or more times each, and K/n words occur n or more times each. Then the number of words occurring exactly n times will be $K/n - K/(n + 1) = K/[n(n + 1)]$.

The rank-size law, often called *Zipf's law* in honor of a U.S. linguist who wrote a great deal about it, holds for just about all of the texts whose vocabularies have been counted, in a great range of languages, not excluding native American languages. But while it holds for *Ulysses,* it fails for Joyce's *Finnegans Wake* (possibly because of the freedom Joyce exercises in creating all sorts of word fragments and variants of dictionary words). A count of ideograms in Chinese texts seemed to show that the law failed; but a recent frequency count of Chinese *words* (each word may consist of one, two, or more ideograms) shows that it fits the Chinese language just as well as it fits others.

[2]In most cases, the first two or three frequencies are substantially lower than the rule predicts, as in Figure 1.

Table 1 The number of rarely occurring words in James Joyce's *Ulysses*

Number of Occurrences (n)	*Number of Words* *Actual*	*Predicted**
1	16,432	14,949
2	4,776	4,983
3	2,194	2,491
4	1,285	1,495
5	906	997
6	637	712
7	483	534
8	371	415
9	298	332
10	222	272

*Predicted number = $K/[n(n + 1)]$; K = 29,899, the total number of *different* words in *Ulysses*.

Why does this regularity hold? Why should the balance between frequent and rare words be exactly the same in a daily newspaper as in Joyce's *Ulysses,* the same in German books as in English books, or the same in most (not all) schizophrenic speech as in normal speech?

Several answers have been proposed, one of which is typical of the explanations that are provided by probability theory. Probability theory often explains the way things are arranged on average by conceding its inability to explain them in exact detail. To explain the laws of gasses it avoids tracing the path of each molecule.

To explain the word distribution, we make some assumptions that might be thought outrageous if applied in detail, but that might be plausible if only applied in the aggregate. We assume that a writer generates a text by drawing from the whole vast store of his or her memory, and by drawing from the even vaster store of the literature of the language. The former of these processes we might call *association,* the latter *imitation.* Specifically, we assume that the chance of any given word being chosen *next* is proportional to the number of times the word has previously been stored away—in memory or in the literature. Remember, these assumptions are intended to apply only in the large. To accept them, we need not believe that Shakespeare wrote sonnets by spinning a roulette wheel any more than we believe the individual molecules of a gas chart their courses by shaking dice.

If we accept the assumptions, then it becomes a straightforward mathematical matter but one beyond the scope of this essay to derive the probability distribution they imply. The derivation yields what is known as the *Yule distribution.* In the upper range, among frequently used words, the Yule distribution agrees with the rank-size law of Zipf; in the lower range, among rarely used words, it gives precisely the observed fractions $1/[n(n + 1)]$.

Now we see why the *same* distribution can fit texts of diverse kinds drawn from the literatures of many languages. The same distribution can fit because it does not depend on any very specific properties of the process that generated the text. It depends only on the generator being, in a probabilistic sense, an associative and imitative process. We might even suspect that substantial departures from exact proportionality in association and imitation would not greatly change the character of the distribution. To the extent that the consequences of changing the assumptions have been explored, mathematically and by computer simulation, the distribution has indeed proved robust. We can give Shakespeare and Joyce a great deal of latitude in the way they write without altering visibly the gross size-rank relation of their vocabularies, but as *Finnegans Wake* shows, we can't give them infinite latitude.

MEGALOPOLIS AND METROPOLIS

Having stripped away some of the mystery of the vocabularies of literary texts, we are perhaps prepared to tackle the corresponding regularity in U.S. city sizes. We have seen (Figure 1) that the city populations obey the same rank-size law, to a quite good approximation.[3] If two cities have ranks j and k, respectively, in the list, their populations' ratio will approximate k/j.

The regularity is not just a happenstance of the 1980 Census. It holds quite well for all the Censuses back to 1780. It does *not* hold, however, for cities in arbitrarily defined geographical regions of the world, which are not relatively self-contained economic units. It does not hold, for example, for Austria, or for individual Central American countries, or for Australia. Nor does it hold if we put the cities of the whole world together (see the uppermost curve in Figure 1). In that case, the distribution is still relatively smooth and regular, but population does not drop off with rank as fast as Zipf's law demands. The distribution is flatter, and the largest metropolises are "too small," though, I hasten to add, this phrase should not be interpreted normatively.

(The definition of size for the world's cities differs from that of the U.S. cities and so the actual magnitudes are not quite comparable. The lists from which they are compiled differ in their year, 1975 vs. 1980, and the notion of metropolitan area may not have been used in deciding the size of a member of the list of world cities.)

In the case of city sizes, then, we must be prepared to explain *two* things: why Zipf's law has held for more than two centuries for the cities of the United States, and why it doesn't hold for many other aggregates of cities. Let's start with the former question and ask what the analogues might be to the association and imitation processes that explained the word distributions. More precisely, let's ask what processes would lead cities to grow at rates proportional, on average, to the sizes already achieved (sometimes called *Gibrat's principle*); for that is the main assumption the mathematical derivation requires.

Cities grow by the net balance of births over deaths, and they grow by the net balance of inward over outward migration. With respect to births and deaths,

[3]We can take either the populations of cities as defined by their corporate boundaries, or populations within metropolitan areas as defined by the U.S. Census. The regularity shows up about as well in either case—perhaps it is a little more satisfactory if we use metropolitan statistical areas.

we need assume only that, on average, birth and death rates are uncorrelated with city size. With respect to migration, we assume that migration outward is proportional, on average, to city size (that is, that per capita *rates* are independent of size), and migration inward (from rural areas, from other cities, or from abroad) is also proportional to city size. The last assumption means that the cities in a given size group form a "target" for migration, which is larger, in total, as the total population already living in the cities of that group is larger. (I leave it to the reader to consider the reasons why this might be a plausible assumption, at least as an approximation.)

If we make these assumptions, we are again led by the mathematics of the matter to Zipf's rank-size law. But now it is instructive to ask: Under what circumstances would we expect a collection of cities to fit the assumptions? The answer is that the cities should form a "natural" region within which there is high and free mobility of population and industry, and which is not an arbitrary slice of a still larger region. The United States fits these requirements quite well, while an area playing a specialized role in a larger economic entity might not fit at all (for example, Austria after the dissolution of the Empire, or a country specializing in agricultural exports and having a single large seaport).

If we put together a large number of distributions, each separately obeying the rank-size law, we get a new distribution of the same shape, simply displaced upward on the graph, but with the top few omitted. We would expect the totality of the world's cities to fit the rank-size distribution, except for a deficiency of extremely large metropolises at the very top, and so it does. If we take the published figures at their face value (the definitional problems are severe, and the census counts of varying accuracy), there are somewhat more than 50 urban aggregations in the world having more than 2 million people each. Zipf's law would then call for a New York or a Tokyo of 100 million people, instead of the mere 20 million who now inhabit each of those cities. But the deficiency of cities at the very top (mostly the top 10) is soon largely made up by the numerous cities of over 5 million population each. Already, the tenth city on the list, Paris, has a population of 9.2 million, only 10% fewer than the number demanded by Zipf's law.

The sizes of cities are of obvious importance to the people who live in them, but it is not obvious what practical conclusions we are to draw from the actual size distribution. One *possible* conclusion is that the distribution isn't going to be easy to change without strong governmental or economic controls over places of residence and work. Or, to put the matter more palatably—because we generally wish to avoid such controls—the mathematical analysis that discloses the forces governing the phenomena teaches us that any attempt to alter the phenomena requires us to deal with those forces with sophistication and intelligence.

BIG AND LITTLE BUSINESS

Economists have generally been more interested in the sizes of business firms than they have been in the sizes of cities. Concentration of industry in the hands of a few large firms is generally thought to be inimical to competition and is generally also supposed to have proceeded at a rapid rate in the United States during the present century.

It has long been known that business firms in the United States, England, and other countries have size distributions that resemble Zipf's rank-size law, except that size decreases less rapidly with rank than in the situations described previously (that is, the ratio of the largest firm to the tenth largest is generally less than ten to one[4]). The slower the decrease in size with increase in rank, the less concentrated is business in the largest firms.

Economists have been puzzled by the fact that the rate of decrease in size with rank, which is one way of measuring industrial concentration, appears to be about the same for large U.S. manufacturing firms at the present time as it was 35 years ago or even at the turn of the century. Even during periods of frequent mergers, the degree of industrial concentration, as measured by the rank-size relation, has changed only slowly.

From our previous analyses, we should be ready to solve the puzzle. Indeed, it can be shown mathematically that under appropriate assumptions about the firms that disappear by merger, and those that grow by merger, mergers will have no effect on concentration. Moreover, the assumptions required for this mathematical derivation fit the United States statistical data on mergers fairly well. In analogy to the processes for words and cities, we can guess what those assumptions—and the data that support them—are like:

- The probability of a firm "dying" by merger should be approximately independent of its size.
- The average assets acquired by surviving firms through mergers should be roughly proportional to the size they have already attained.

And these are indeed not very far from the truth.

Thus a line of scientific inquiry that began with a linguistic puzzle over word frequencies leads to an explanation of a paradox about industrial concentration in the United States. That explanation opens new lines of research for understanding business growth and arriving at public policy for the maintenance of business competition.

Our fascination with rank-size distributions need not stop with the three examples examined here. We may expect the Zipf distribution to show up in other places as well, and each new occurrence challenges us to formulate plausible (and testable) assumptions from which the rank-size law can be derived and

[4]Let m and n be the ranks of two members of a rank-size distribution, and let S_m and S_n be their respective sizes. Then the rank-size law, in this generalized form, requires $S_m/S_n = (n/m)^k$, where the exponent k is a proper fraction. When k approaches unity as a limit, we get the special case of the Zipf distribution. The general distribution is usually called by economists the ***Pareto distribution.*** If we graph the logarithmic distribution, taking logarithms of both ranks and sizes, we again obtain a straight line with a slope equal to the fraction k. The steeper this straight line (the larger k), the larger are the first-ranking firms compared with the firms further down the list (that is, the larger is k, the greater the concentration).

the occurence explained. I will leave a final example as an exercise for the reader. List the authors who have contributed to a scientific journal over a span of 20 years, or whose names have appeared in a comprehensive bibliography, such as *Chemical Abstracts.* Note the number of appearances for each author, and rank the authors by that number. Then about one-half of all the authors will have appeared exactly once, one-sixth will have appeared twice, and so on; the data will not stray far from the Yule distribution. What are the ways of authors that can provide a naturalistic explanation for that fact?

PROBLEMS

1. From Figure 1, what roughly is the population of Philadelphia, the fifth largest city in the United States?
2. Consider Table 1.
   a. 483 distinct words appear 7 times in the text of *Ulysses.* How was the predicted value of 534 computed?
   b. Suppose the predicted number of words occurring n times is 164. Approximately what is n? (Hint: You will have to use the quadratic formula from high school algebra.)
3. What does the author mean by *association*? By *imitation*?
4. How can the same distribution fit the population of U.S. cities and the frequency of words in a text?
5. What is *Gibrat's principle*? How does it relate to Zipf's law? To the sizes of U.S. cities?
6. State in words the mathematical assumptions that lead to the Yule distribution, first in the case of literary texts, then in the case of city sizes.
7. Can you think of a reason why the few largest cities in the United States might not satisfy the rank-size law?
8. a. How are the Zipf and Pareto distributions related?
   b. In a logarithmic graph of the Pareto distribution, what is k?
9. Suppose IBM's and Xerox's sales rank first and fourth, respectively, Xerox's sales are $1 billion, and business-firm sales obey a Pareto distribution with $k = 1/2$. What would IBM's sales be?
10. If R_n is the rank of a thing of size n, state Zipf's law.
11. Assume the fourth largest city has a population of 10 million. What ranking would you expect a city of 2½ million to hold, according to Zipf's law?
12. The author states that the number of different words occuring exactly n times in a given text equals $K/[n(n + 1)]$. What is K? (Hint: See Table 1.)

13. Refer to the exercise outlined in the final paragraph. What two assumptions would you postulate for this distribution? (Hint: Use the assumptions for mergers or city sizes as a close guide.)

REFERENCE

Yuji Ijiri and Herbert A. Simon. 1977. *Skew Distributions and the Sizes of Business Firms.* Amsterdam: North-Holland Publishing Company.

How Accountants Save Money By Sampling

John Neter *University of Georgia*

Accountancy and statistics are regarded by many people as two of the dullest subjects on earth. The essays in this volume, it is hoped, will change such views about statistics. This essay deals with important uses of statistics in accounting practice, and it may also reveal some interesting facets of accounting.

All of us, after all, want to use our money efficiently and effectively. We shall see how the use of statistical sampling in accounting saves money for railroads and airlines as they face problems of dividing revenues among several carriers. Similar statistical sampling methods are used in other areas of accounting and auditing work. Indeed, they are used in many fields of business, government, and science.

Accountants and auditors traditionally have insisted on accuracy in the accounting records of firms and other organizations. This insistence has led them to do much work on a complete, 100% basis. For instance, auditors may want to check the value of the inventory that a firm has on hand. To do this, they may examine the entire inventory; that is, they may actually count how many units of each type of inventory item are on hand, determine the value of each kind of unit, and thus finally obtain the total value of the inventory.

As another instance, an auditor may want to know the proportion of accounts receivable that have been owed for 60 or more days. This information may be needed to verify a reserve for bad debts. Accounts that have not been paid within 60 days are more susceptible to bad debt losses than accounts that have been open a shorter time. In order to establish the proportion of accounts receivable that are 60 days old or older, the auditor may examine every single account receivable held by the firm and determine for each the amount of money owed for 60 or more days.

Is it necessary to conduct these 100% examinations of inventory, of accounts receivable, or of similar collections, in order to obtain the figures that the accountant needs? More specifically, could a sample adequately provide the information needed by the accountant without all of the tedious work necessary for a complete, 100% examination? Let us focus on the inventory items.

In statistical terminology, the group of inventory items for which the total value is to be ascertained is called the ***population*** of interest. A ***sample*** selected from such a population consists of some, but not all, of the items in the population. A sample is selected to find out about characteristics of the population without looking at every element of the population.

The cost of examining a relatively small sample of inventory items is usually less than the cost of a complete examination because the sample requires an examination of fewer items. But are the results based on the relatively small sample almost as good as those from a complete examination?

Experience with sampling in many areas has shown that relatively small samples frequently provide results that are almost as good as those obtained from a complete examination, while at the same time the sample results cost considerably less. Indeed, sometimes the sampling results are even better than those from a 100% examination. That statement may seem startling, but consider the task of taking an inventory in a large company. Many persons are required for the task. Because of the size of the undertaking, it may be hard to give thorough training to these persons, and the quality check on the work may have to be limited. On the other hand, a small sample of the inventory items would require fewer persons, and therefore they could be trained better. Furthermore, the quality control program for the inventory could be more rigorous when a smaller number of persons are involved. The net effect might well be that the sample results are more accurate than the 100% enumeration! That is, the gains in accuracy from better training and quality control with a small sample may more than balance the sampling error introduced by selecting only a sample of inventory items instead of all of them. Of course, the sampling must be done intelligently and properly. The study of sampling is an important part of statistics.

THE CHESAPEAKE AND OHIO FREIGHT STUDY

Statements that relatively small samples can provide results almost as good as those from a complete examination, or indeed sometimes even better, are often not convincing by themselves. Statisticians have therefore often found it helpful to conduct studies that compare the results of a sample with those of a complete enumeration. Such a study was made by the Chesapeake and Ohio Railroad Company in determining the amount of revenue due them on interline, less-than-carload freight shipments. If a shipment travels over several railroads, the total freight charge is divided among them. The computations necessary to determine each railroad's revenue are cumbersome and expensive. Hence the Chesapeake and Ohio experimented to determine if the division of total revenue among several railroads could be made accurately on the basis of a sample and at a substantial saving in clerical expense.

In one of these experiments, they studied the division of revenue for all less-than-carload freight shipments traveling over the Pere Marquette district of the Chesapeake and Ohio and another railroad (to be called A for confidentiality), during a six-month period. The waybills of these shipments constituted the population under examination. A waybill, a document issued with every shipment of freight, gives details about the goods, route, and charges. From it, the amounts due each railroad can be computed. The total number of waybills in the population was known, as well as the total freight revenue accounted for by the population of waybills. The problem was to determine how much of this total revenue belonged to the Chesapeake and Ohio.

For the six-month perod under study, there were nearly 23,000 waybills in the population. Since the amounts of the freight charges on these waybills vary greatly (some freight charges were as low as $2, others as high as $200), it was decided to use a sampling procedure called ***stratified sampling.*** With this procedure, the waybills in the population are first divided into relatively homogeneous subgroups called ***strata.*** The subgroups in this instance were set up according to the amount of the total freight charge, since the amount due the Chesapeake and Ohio on a waybill tended to be related to the total amount of the waybill. That is, the larger the total amount of a waybill, the larger tended to be the amount due the Chesapeake and Ohio on that waybill. Specifically, the strata were as follows:

Stratum	*Waybills with Charges Between*
1	$ 0 and $ 5.00
2	$ 5.01 and $10.00
3	$10.01 and $20.00
4	$20.01 and $40.00
5	$40.01 and over

Note that each stratum contains waybills with total freight charges of roughly the same order of magnitude. Because of the general tendency by which the amount due the Chesapeake and Ohio varied with the total freight charge on a waybill, each stratum is relatively more homogeneous with respect to the amount of freight charges due the Chesapeake and Ohio. At the same time, the strata differ substantially from one another.

Statistical theory then was used to decide how large a sample from each stratum must be selected so that the amount of the revenue due the Chesapeake and Ohio could be estimated with required precision from as small a sample as possible. One piece of information needed for this determination was the number of waybills in each stratum. The sampling rates decided on for the strata were:

Stratum	*Proportion to Be Sampled*
1	1%
2	10%
3	20%
4	50%
5	100%

Note that this theory led to larger sampling rates in the strata containing wider ranges of freight charges and smaller sampling rates in the strata containing narrow ranges of freight charges. To understand this, consider stratum 1, containing waybills with charges between $0 and $5.00. Here the variation between the waybill amounts is small, and therefore a small sample will provide adequate information about the amounts of all of the waybills in that stratum. On the other hand, stratum 4, containing waybills with charges between $20.01 and $40.00, has much greater variation. A larger sample is therefore required in this stratum to obtain adequate information about the amounts of all waybills in that stratum. In an unreal extreme situation where all the waybills in a stratum would have the same amount due the Chesapeake and Ohio, a sample of just one waybill would provide all the information about the waybill amounts in that stratum.

Once the sample sizes were determined, the next problem was to select the samples from each stratum. For a statistician to be able to evaluate the precision of the sample results (that is, how close the sample results are likely to be to the relevant population characteristic), the sample must be selected according to a known probability mechanism. Various methods of probability sampling are available. One is called *simple random sampling.* This type of sample may be directly selected by use of a random number generator or by use of a table of random numbers, a portion of which is illustrated in Table 1. How might Table 1 be used to select a simple random sample from each of the strata? Consider stratum 1 and suppose it contains 9,000 waybills, which we label with four-digit numbers from 0001 to 9000. We want to obtain four-digit numbers from the table; we might start in the upper left-hand corner, using columns 1 through 4. The first number obtained is 1328. Our first sample waybill is then the one numbered 1,328. Our second sample waybill would be 2,122. The next number from the table of random digits is 9905, but there

Table 1 Portion of a table of random digits

Line	*(1)–(5)*	*(6)–(10)*	*(11)–(15)*	*(16)–(20)*	*(21)–(25)*	*(26)–(30)*	*(31)–(35)*
101	13284	16834	74151	92027	24670	36665	00770
102	21224	00370	30420	03883	94648	89428	41583
103	99052	47887	81085	64933	66279	80432	65793
104	00199	50993	98603	38452	87890	94624	69721
105	60578	06483	28733	37867	07936	98710	98539
106	91240	18312	17441	01929	18163	69201	31211
107	97458	14229	12063	59611	32249	90466	33216
108	35249	38646	34475	72417	60514	69257	12489
109	38980	46600	11759	11900	46743	27860	77940
110	10750	52745	38749	87365	58959	53731	89295
111	36247	27850	73958	20673	37800	63835	71051
112	70994	66986	99744	72438	01174	42159	11392
113	99638	94702	11463	18148	81386	80431	90628
114	72055	15774	43857	99805	10419	76939	25993
115	24038	65541	85788	55835	38835	59399	13790
116	74976	14631	35908	28221	39470	91548	12854
117	35553	71628	70189	26436	63407	91178	90348
118	35676	12797	51434	82976	42010	26344	92920
119	74815	67523	72985	23183	02446	63594	98924
120	45246	88048	65173	50989	91060	89894	36036

Source: Table of 105,000 Random Decimal Digits. Interstate Commerce Commission, Bureau of Transport Economics and Statistics, Washington, D.C., May 1949.

are only 9,000 waybills in the stratum, so we pass over this number and go on to the next one, which is 0019. This process would be continued until the required sample of 90 (1% of 9,000) has been obtained. The digits in the table of random digits are generated so that all numbers (four-digit numbers in our case) are equally likely.

Another method of selecting waybills from each stratum is called *serial number sampling,* and this was the method actually used by the Chesapeake and Ohio Railroad. In this procedure, the sample within each stratum is selected according to certain digits in the serial number of the waybill. In this particular case, the last two digits in the serial number of the waybill were used. To explain how these last two digits are used to select the sample, consider stratum 1, with its 1% sample. The number of possible pairs of digits appearing in the last two places of the serial number (00, 01, 02, . . . , 99) is 100. If one of these pairs is chosen from a table of random digits and all waybills with these last two digits in their serial number are selected for the sample, it will be found that about 1% of the stratum is included in the sample. For stratum 1, the random number turned out to be 02. Therefore all waybills with freight charges of $5 or less whose last two serial number digits were 02 were selected for the sample. The serial number digits used for the other strata, as well as the sampling rates, were as follows:

Stratum	*Proportion to be Sampled*	*Waybills with Numbers Ending in:*
1	1%	02
2	10%	2
3	20%	2 or 4
4	50%	00 to 49
5	100%	00 to 99

Since the serial numbers appear prominently on the waybills, this procedure is a simple one for selecting the sample. Furthermore, in this case experience indicates that it provides essentially the equivalent of a simple random sample from each stratum.

Altogether, 2,072 waybills out of 22,984 in the population (9%) were chosen according to this procedure. For each waybill in the sample, the amount of freight revenue due the Chesapeake and Ohio was calculated. For each stratum, the total amount due for the population of waybills was then estimated, and these estimates were added to obtain an estimate of the total amount of freight revenue due the Chesapeake and Ohio on the almost 23,000 waybills in the population. Because this was an experiment, a complete examination of the population was also made, so that the sample result could be compared with the result obtained from an analysis of all waybills in the population. The findings were:

Total amount due Chesapeake and Ohio on basis of complete examination of population	$64,651
Total amount due Chesapeake and Ohio on basis of sample	64,568
Difference	$ 83

Thus a sample of only about 9% of the waybills provided an estimate of the total revenue due the Chesapeake and Ohio within $83 of the figure obtained from a complete examination of all waybills. Because the sample cost no more than $1,000, while the complete examination cost about $5,000, the advantages of sampling are apparent. It just does not make sense to spend $4,000 to catch an error of $83. Furthermore, although the error in this instance was against the Chesapeake and Ohio, the next time it may be against another railroad, so that the long run cumulative error is relatively even smaller.

OTHER RAILROAD AND AIRLINE SAMPLING STUDIES

The Chesapeake and Ohio conducted the same type of test for interline passenger receipts. They studied tickets sold during a five-month period to commercial passengers traveling only on the Chesapeake district of the Chesapeake and Ohio and on two other railroads, A and B. The findings are shown in Table 2. Again, these results dramatically demonstrate the ability of relatively small samples to provide precise estimates of the total revenue due the Chesapeake and Ohio.

Airlines also have used statistical sampling to estimate their share of the revenue on tickets for passengers traveling on two or more airlines. Three airlines tested statistical sampling during a four-month period. In that time, the degree of error in the sample estimate based on relatively small samples did not exceed 0.07% (that is, $700 in $1,000,000) for any of the three airlines. On the basis of this experiment, wider use of statistical sampling in settling interline accounts has been made. At one point in time, the sample consisted of about 12% of the interline tickets and the cumulative sampling error was running at less than 0.1%. The clerical savings were estimated to be near $75,000 annually for some of the larger carriers and more than $500,000 for the industry.

Statistical sampling in accounting and auditing has also been used to estimate the value of inventory on hand, the proportion of accounts receivable balances

Table 2 Results of passenger ticket study

	100% Examination	*5% Sample*	*Difference Dollars*	*Difference Percent*
Railroad A				
Total number of tickets	14,109			
Total revenue	$325,600			
Chesapeake and Ohio portion of total revenue	$212,164	$212,063	−$101	−0.05
Railroad B				
Total number of tickets	7,652			
Total revenue	$128,503			
Chesapeake and Ohio portion of total revenue	$ 79,710	$ 80,057	+$347	+0.44

Source: Railway Age, June 9, 1952.

that are 60 days old or older, and the proportion of accounts receivable balances that are acknowledged as correct by the customer. In each instance, it has been demonstrated that a relatively small sample, carefully drawn and examined, can furnish results that are of high quality and at a much lower cost than with a complete examination.

SAMPLING FOR RADIO ROYALTIES

Another area where statistical sampling is used for accounting purposes determines the distribution of royalties to composers and publishers for music played on the radio. A performing rights organization, BMI, collects fees that entitle radio stations to use BMI's affiliates' music. Each station pays a blanket fee, proportional to the revenue of the station. These fees then must be apportioned appropriately to the composers and publishers of the music actually played. Information on the music played on the radio stations cannot be obtained on a 100% census basis because it would be much too costly. There are around 9,000 stations playing music an average of 18 hours per day. Thus, for 365 days in a year, these stations play approximately 59 million hours of music.

Instead, BMI selects a sample of radio stations to determine what music is being played. It does so by first stratifying all of the stations in the United States according to urbanity, region of the country, and amount of music played per day. This is done by a computerized program that also groups the stations according to their similarity with respect to the proportion of BMI music that they play. This stratification is made annually and usually leads to eight or nine strata. Each stratum is then divided into two substrata according to the size of the fee paid to BMI, so that stations that pay small fees and stations that pay large fees are placed into different strata. In addition to these strata, another stratum is established for new stations that go into operation during the year.

A sample of radio stations is then randomly selected each quarter from each stratum. For each selected station, two or three days during the quarter are randomly chosen, with the days distributed uniformly over the quarter. During these days, each selected station keeps a log of every song played. These logs are then returned to BMI, the information is entered into the computer, and the music is identified either as music for which BMI is the representative or music for which BMI is not the representative. From this sample information, projections are made of the amount of payments that BMI should make to each music publisher and composer each quarter. Over the course of a year all radio stations are asked to log one specified period.

SUMMARY

To summarize, statistical sampling consists of the selection of a number of items from a population, with the selection done in such a way that every possible sample from the population has a known probability of being chosen. Frequently, a statistical sample can provide reliable information at much lower cost than a complete examination. Also, a statistical sample often can provide more timely data than a complete enumeration of the population because fewer data have to be collected and smaller amounts of data need to be processed. Finally, a statistical sample can sometimes provide more accurate information than a complete enumeration when quality control over the data collection can be carried on more effectively on a small scale.

PROBLEMS

1. Explain the difference between simple random sampling and serial number sampling.

2. Suppose a university administrator is considering ordering some new desks for classrooms. She needs to find out how many desks already in use need to be replaced.
   a. Should she consider using sampling methods in this situation? What are the arguments for sampling? Against?
   b. If she did use sampling methods, what would the *population* be?

3. Why was stratified sampling used in the C&O freight study?

4. Refer to Table 2. Add the Railroad A and Railroad B ticket revenues, and find the difference in percent between a 5% sample and a 100% examination.

5. In the C&O freight study, how large a percentage of the total amount due C&O was the result of error due to sampling?

6. An army psychologist wants to take a sample of 1,000 enlisted men to find out their attitudes toward the "new Army." He obtains a list of the names of 10,000 enlisted men arranged by squads; each squad has 10 men, with a sergeant heading the list, then a corporal, followed by 8 privates.
   a. Would you recommend that the psychologist use serial number sampling (using the digits 0–9) to choose a sample of 1,000 from this list of 10,000 men? Why?
   b. If the psychologist used serial number sampling, what would be the chance of getting only sergeants in his sample? What if he used simple random sampling?
   c. Answer the questions in (a) and (b) if the psychologist used a list that placed the 10,000 enlisted men in alphabetical order.

7. Suppose C&O and Railroad A sampled tickets to determine their share of revenue from interline passenger receipts every month for a year. For how many months would you expect the sampling error to favor C&O?

8. Use Table 1 to draw a random sample of 25 two-digit numbers. How many are even? How many have both digits even? Do the same for a random sample of 100 two-digit numbers. Compare your answers to those obtained by others. What conclusions can you draw?

REFERENCES

A. A. Arens and J. K. Loebbecke. 1981. *Applications of Statistical Sampling to Auditing.* Englewood Cliffs, N.J.: Prentice-Hall.

"Can Scientific Sampling Techniques Be Used in Railroad Accounting?" *Railway Age,* June 9, 1952, pp. 61–64.

D. A. Leslie, A. D. Teitlebaum, and R. J. Anderson. 1979. *Dollar-unit Sampling: A Practical Guide for Auditors.* Toronto: Copp Clark Pitman.

T. W. McRae. 1974. *Statistical Sampling for Audit and Control.* London: Wiley.

M. S. Newman. 1982. *Accounting Estimates By Computer Sampling,* 2nd ed. New York: Wiley.

T. M. F. Smith. 1976. *Statistical Sampling for Accountants.* London: Accountancy Age Books.

A. J. Wilburn. 1984. *Practical Statistical Sampling for Auditors.* New York: Marcel Dekker.

Preliminary Evaluation of a New Food Product*

Elisabeth Street
Mavis B. Carroll

Many Americans like to have at hand an easy-to-prepare, nutritious, on-the-run meal. Our company has been developing such a product, called H.

Development of such a product calls for a thorough evaluation. In this essay we limit ourselves to two aspects of the evaluation: the protein content of H and its tastiness. We shall show how statistics was central in answering our questions.

The palatability of a product can be determined by having people taste the prepared product and evaluate its acceptability both overall and by specific attributes. The nutritive quality of a food product can be determined by feeding it to animals whose metabolic processes are very similar to ours. Both types

*This article is reprinted from the first edition (1972). It was written when the authors were on the staff of the General Foods Corporation and reflects the situation as it was at that time. The statistical methods used continue to be current in such investigations, though evolutionary improvements have also come into use. *The editors.*

of tests should be designed, that is, planned in detail in advance. This helps avoid a consistent error in one direction (bias) and ensures that the proper number of people and animals are included in each study to answer with a fair degree of confidence the questions being posed. Some of the steps in the design and analysis of such tests are described below.

PROTEIN EVALUATION

The protein content of H, in one sense, was satisfactory, as calculations based on the constituents of H showed. We were not certain, however, about the efficiency of the protein in H under conditions of actual use, and, in addition, we wanted to compare two forms of H, one solid and the other in liquid form.

A rat-feeding study can give practical support to the high-protein claim and a way of comparing the two variants of H because the rats' responses would be affected by any interaction of the ingredients in the formula or by a shortage of an essential amino acid making the protein less efficient. Neither of these conditions would be indicated by paper calculation.

Previous experience had shown that 28 days of feeding a diet to 10 to 15 rats gives a fairly reliable estimate of the diet's protein efficiency. This feeding study, as do all such studies, used male rats, newly weaned. Male rats grow faster than female rats, and while weanlings, they are in a period of maximum growth rate. Adult rats are not used, for their weights are stabilized, and it is the animals' weight gain that is of primary interest.

Besides comparing the two forms of H, the experiment also compared each with a casein control diet, which served two purposes. First, it is a standard diet to which many experimental diets are compared; second, because we have had much experience with the casein diet, it would provide a check on whether something was amiss with the batch of rats or with some other aspect of the study. Things sometimes go wrong in mysterious ways, and it is important to have some sort of check.

At the start of the experiment, the 30 animals used varied in weight from 50 to 63 grams. They were arranged in ascending order of weight, and from the 3 lightest ones, one was assigned to each of the three diets in the assay. Such a trio of approximately equal weight animals is called a *block*. The assignment of rats to diets within this block of 3 was by chance or at random; that is, random numbers determined which rat went on each diet. The second block of the next 3 lightest rats was assigned one to each diet in the same way. This process of randomized block assignment continued until all 30 rats set aside for this study were distributed among the three diets, 10 per diet.

Using blocks of rats of comparable body weight ensures that each diet has its proper share of light and heavy rats. Randomization helps balance other factors that might influence the outcome of the study. For example, the shipment of 30 rats probably included many who were littermates. The rats were all newly weaned and so were all born at approximately the same time. The probability is high that the randomization spread the rats from one litter among the three diets. Of course, to be sure that a litter is equally divided among the three diets it would be necessary to identify the rats by litter, reduce a litter size to a multiple of three by random withdrawal of animals, and randomly assign the remaining rats equally to the three diets. This would be done if some inheritable trait might importantly affect the results of a study.

The rats were assigned to cages by random numbers because prior studies had shown that rats at the top of a rack of cages gain more weight than those at the bottom. This randomization guarded against the possibility that most of the rats on one diet were put in top cages while rats on another diet went into bottom cages.

During the H feeding study each rat was permitted to eat as much as he wished. His food consumption was measured, and his weight was checked weekly. The final weight gain was simply the difference between his final, 28-day body weight and his initial weight. Intermediate weights were used to check on any untoward event such as respiratory infection because, in the rat, body weight is a sensitive indicator of general well-being. The total food consumed times the proportion of protein—a value obtained by chemical analysis of samples from each diet—gives a rat's protein intake. For the H study, the results for these two measures are shown in Figure 1. Each point on the graph represents one rat. All three diets were made up to have about the same proportion of protein, 9%. The chemical analysis showed them to be close to that: 9.875%, 9.875%, and 9.50%. Therefore the higher intake of protein for rats fed the H diets means they ate more food, which may indicate that these diets were more palatable to the rats than the casein diet. But experienced nutritionists claim that rats generally will eat heartily any balanced high-protein food irrespective of texture and flavor. All that is known from the data is that the intake of H was substantially above the intake of casein. The other most obvious

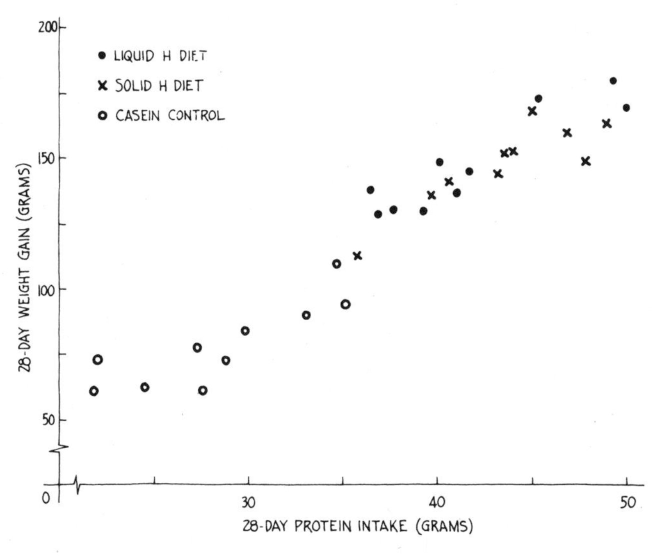

Figure 1 *Relationship of 28-day protein intake and weight gain in young male rats.*

fact about the points in Figure 1 is that they tend to fall close to a slanted straight line drawn through the middle of them. But closer inspection of Figure 1 with ruler in hand indicates that a freehand straight line through the liquid H points lies slightly above a line through solid H points and has a steeper slope. A line through the casein points alone would lie below either of the H lines, if the casein line were extended to the higher intakes. Thus some doubt was cast on the initial conclusion that all points fell on one line and thus on the hypothesis that any differences between the dietary weight gains could be attributed entirely to differences in protein intake; that is, we began to think that the difference in weight gain was not due simply to the difference in protein intake.

We used a computer program to fit a straight line to the points for each diet by the method of least squares; that is, the line was selected that minimized the sum of the squared vertical distances of the points from the line. These fitted lines, called *estimated regression lines,* are described by equations of the form $Y = \bar{Y} + b(X - \bar{X})$, where Y is the predicted weight gain and X the protein intake. $\bar{X}$ and $\bar{Y}$ are the averages of the Xs and Ys separately for each set of data, and each b represents the slope of that straight line, that is, the predicted increment in weight gain per unit of protein intake.

The three regression equations were:

$$\begin{aligned} &\text{Liquid H:} && Y = 151 + 3.72(X - 41.7) \\ &\text{Solid H:} && Y = 150 + 3.66(X - 43.3) \\ &\text{Casein:} && Y = 79 + 2.91(X - 28.4). \end{aligned}$$

Inspection of the three lines—they are graphed in Figure 2—shows that the casein line is decidedly lower and slopes less than the two H lines. The two H lines were then compared, and we found that they did not differ more than one would expect by chance. Thus our suspicion was confirmed that the greater gains of the H-fed rats relative to the gains for rats on casein were not due simply to greater protein intake because the slope of the H lines is higher.

To summarize the results of the protein evaluation: differences in protein intake between the casein-fed rats and H-fed rats did not account for all the differences in weight gain. For a given increase in protein intake the H diets resulted in a greater increase in weight gain than did the casein diet. There was little difference between solid and liquid H in terms of estimated weight gain. If the addition of solidifying agents is detrimental it was not evident in this assay.

PALATABILITY EVALUATION

While the study to determine and compare the nutritive values of the two H variants was being run, a preliminary consumer test to determine the palatability of the two variants was also run. The aim was to compare the two H products and to see if they were as acceptable overall as a competitive product already on the market, designated C.

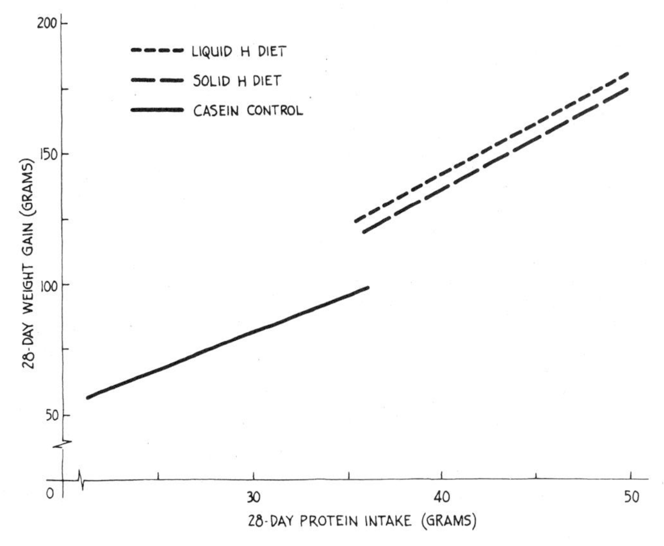

Figure 2 *Estimated regression of 28-day weight gain on protein intake for young male rats.*

Previous experience had shown that having 50 people taste and evaluate a product under controlled conditions is adequate to reveal a major problem in acceptability, if one exists. Controlled conditions were obtained by bringing individuals into a central testing location and having trained personnel prepare, serve, and interview the 150 individuals required—50 to taste C and 50 to taste each of the two variants of H. Because men and women might respond differently, the test specified that each product would be tasted by 25 males and 25 females. The tasters, who were paid for their help, were recruited from local churches or club groups. As they were recruited, individuals were assigned randomly to taste each of the three products until 25 of each sex had tasted and evaluated each of the three products.

After tasting the product, each person was asked to mark a ballot rating overall acceptability. Figure 3 shows the pictorial scale used to measure acceptability. (Individuals seem to find it easier to express their feelings about a product using a pictorial scale rather than a scale using words such as *excellent, very good, good, average, poor, very poor,* and *terrible.*) Each of the pictures is assigned a number (or score) in the sequence +3, +2, +1, 0, −1, −2, −3, with +3 being most acceptable.

For each of the three products the total number of votes for each of the pictures was tallied separately for males and females. The results, called *frequency distributions,* are shown in Table 1 with the pictures being replaced by the assigned number or score, and M indicating males and F females.

From the frequency distributions it is apparent that all those tasting the same product did not agree on its acceptability. It is difficult to look at the six distributions and decide whether any one group of 25 is scoring what they tasted as more or less acceptable than each of the other five groups scored what they

Figure 3 *Pictorial taste-test ballot (the scores +3 to −3 were assigned the figures from left to right).*

tasted. To simplify the comparison among the six groups of 25 people, an average score is obtained by multiplying each score by its frequency, summing the results, and dividing by 25.

We note that the differences in average scores are not large when male tasters are compared to female tasters. We note too that the averages for liquid H and solid H don't differ much, but all four of the H averages are well above the two C averages. The question to be answered is whether these differences in average scores are larger than can be expected by chance, considering the way people vary when they rate the same product. What is needed is a yardstick to permit us to say what difference between any two averages, each based on 25 tasters, is larger than can be expected by chance alone.

To arrive at such a yardstick we must first measure how variable people within each of the six groups are. This measure, called *variance,* is obtained by taking the difference of each person's score from the average, squaring the differences, summing the squares, and dividing by one less than the number of people. For each of the six groups the variance was computed; it is shown at the bottom of the table of frequency distributions. The variances for all six groups in this study were then averaged to give 1.32, a good measure of variability among males or females tasting the same product.

Table 1 Frequency distribution of scores

	Product Tasted					
	C		*Liquid H*		*Solid H*	
Score	*M*	*F*	*M*	*F*	*M*	*F*
+3	1	0	4	3	2	3
+2	2	3	6	7	6	5
+1	7	8	9	7	10	9
0	8	9	5	6	6	7
−1	5	3	0	2	1	0
−2	2	1	1	0	0	1
−3	0	1	0	0	0	0
Total tasters	25	25	25	25	25	25
Average score	0.20	0.24	1.24	1.12	1.08	1.04
Variance	1.50	1.43	1.44	1.36	0.993	1.20

If individuals vary, so will the averages based on individuals. How much the averages vary will depend on the number of tasters: the larger the number the more representative and less variable the average will be. Because of this variability we can never be absolutely sure that two averages represent real differences, but we must take some risk in drawing conclusions. So the best we can do is state the risk in setting up our yardstick. We chose to take a "1-in-20" chance of being wrong when we calculated the amount by which two averages had to differ in order for us to conclude they represent real differences. Using the variance of the average of 25 scores and taking into account the risk gives 0.64 as our yardstick for comparing any two of the average scores.

This tells us that there is no evidence that males and females who rate the same product rate them differently because we have for the differences in averages the following:

$$\begin{aligned} &\text{C:} && 0.24 - 0.20 = 0.04 \\ &\text{Liquid H:} && 1.24 - 1.12 = 0.12 \\ &\text{Solid H:} && 1.08 - 1.04 = 0.04. \end{aligned}$$

We can now combine the distributions for males and females. The resulting distributions are shown in Figure 4. We note that the distribution for product C is shifted to the right, the less acceptable scores, as compared to the H results. There is considerable overlapping of the distributions representing the variation, so even on a pictorial basis we have problems interpreting the relative acceptability of the test products. We must resort to the average scores, which for the 50 individuals who tasted the same product are the following:

$$C = 0.22 \qquad \text{Liquid H} = 1.18 \qquad \text{Solid H} = 1.06.$$

Our yardstick for judging differences decreases because these new average scores are based on 50 people. Our new yardstick, taking a 1-in-20 chance of being wrong, is 0.45.

It is apparent the difference between the acceptability of the two H products is not large enough to say one is more acceptable than the other. There is very little risk, however, in concluding that both H products are more acceptable than C, the competitive product.

Thus, through testing and with the application of statistical methodology, these new products were shown to be palatable and to live up to the concept of a high-protein food. Two of the early criteria in the long process of introducing a new food product have been met.

PROBLEMS

1. For the protein evaluation, why were rats chosen as the experimental animals?

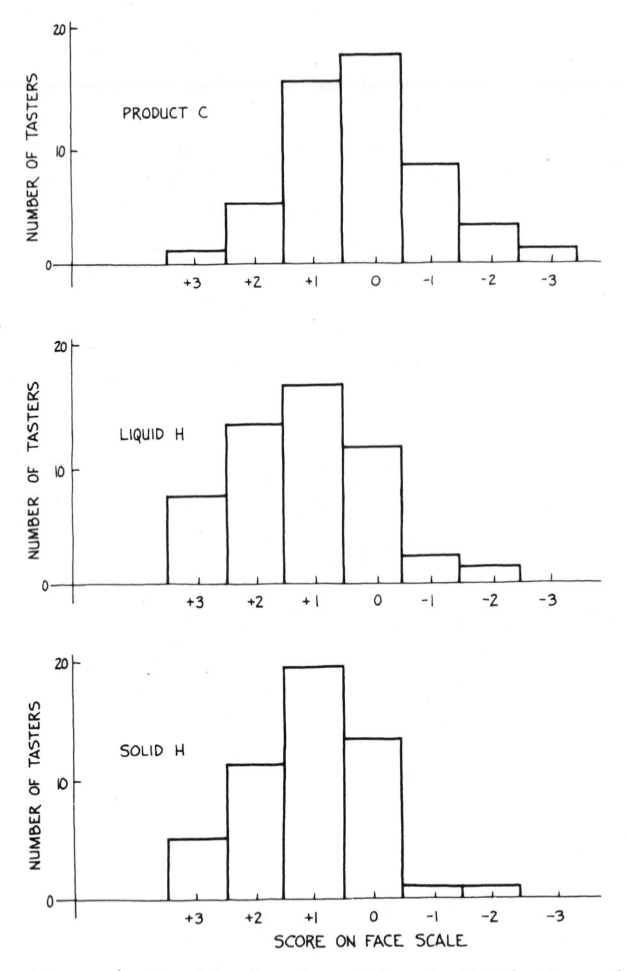

Figure 4 *Combined male and female distributions of face-scale scores by product tasted, 50 tasters per product.*

2. a. Explain the idea of blocking.
   b. Suppose the rats had not been assigned via blocking but instead were assigned in the following way: For each of the 30 rats a six-sided die is rolled. A 1 or a 2 means the casein diet for that rat, 3 or 4 liquid H, and 5 or 6 the solid H diet. Under this scheme how many rats would be assigned to each diet? Would it be possible for all rats to be assigned to the casein diet?

3. In the protein evaluation experiment, what was the least protein intake for any rat on the casein diet? The most? Answer the same questions for the liquid H diet and the solid H diet.

4. What was median weight gain of the 30 rats in the protein-evaluation experiment?

5. Using the appropriate equation derived from the protein-evaluation experiment, how much weight would you predict a rat would gain in 28 days if he consumed 45 grams of protein on the casein diet? State the assumptions that led you to this prediction.

6. Suppose you were setting up the protein-evaluation experiment. Explain how you would use the random number table on page 155 to
   a. Assign rats in each block to the three different diets.
   b. Assign rats to the top or bottom cages.

7. a. The yardstick for comparing average scores between males and females tasting a food in the palatibility evaluation was 0.64; but for comparing the average scores between the two different food products, it was 0.45. Why are these yardsticks different?
   b. Do you think you could ever achieve a yardstick smaller than 0.1? If your answer is yes, how might you do this?

8. Calculate the variance of the following scores for a product B.

	Score	*Frequency of Score*
	+3	2
	+2	7
	+1	7
	0	7
	−1	2
	−2	0
	−3	0
Total Tasters		25
Average Score		1

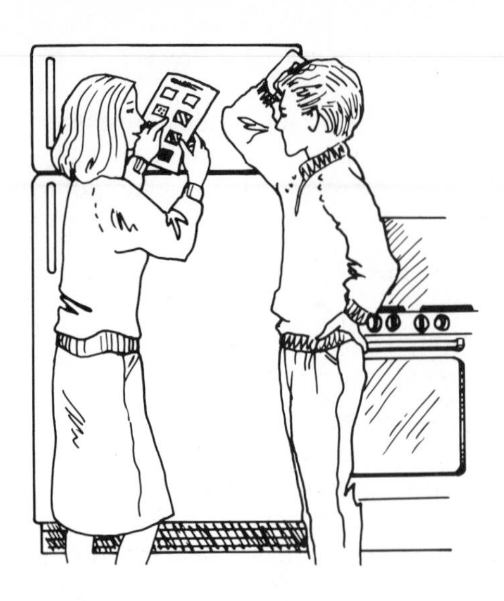

Statistical Determination of Numerical Color Tolerances

Lonnie C. Vance *Cadillac Motor Car Division, General Motors Corporation, Detroit, Michigan*

INTRODUCTION

After 10 years of faithful service, a couple's refrigerator gives out. Their other major kitchen appliances, namely, the electric range and dishwasher, are still operating satisfactorily. All three appliances were purchased at the same time, are of the same make, and match in color. Although their appliances are 10 years old, the couple is happy to discover when shopping for a new refrigerator that the manufacturer of their original appliances makes a refrigerator that they like and that seems to be the same color as the original. Only one problem exists. The "same" color for the new refrigerator does not exactly match the color of their range and dishwasher. They are faced with several alternatives—all of which are less than desirable. Do they replace their range and dishwasher with new ones that will match the color of their new choice for a refrigerator? Do they buy the new refrigerator and live with the slight mismatch of colors among their appliances? Do they have their present range and dishwasher painted to match, if possible, the color of their new refrigerator?

The problem of color matching is not new. Many industries, such as paint, dye, textile, and automotive, have long dealt with color matching. As an illustration, the army wants its olive-green uniforms to have coats and trousers that match in color, and to have recently purchased uniforms match previous ones.

The appearance of the interior and exterior surfaces of an automobile is important to a potential car buyer. Without a good color match between parts of the "same" color, the customer is dissatisfied. For example, the glove box and ashtray doors may be part of the instrument panel. One obviously wants the various parts of the instrument panel to match. But these various components of the "same" color and same material might be manufactured at different locations and later assembled side by side in the automobile's interior. How can an automobile manufacturer have some assurance that such parts when assembled side by side will match in color?

Similarly, the customer wants a good color match among the various painted surfaces on the exterior of an automobile. Some of these adjoining surfaces may be made of different materials and processed differently.

Historically, colors have been matched subjectively by means of individual visual observations. But because of the critical nature of the matches, the various materials and parts that must be matched, and our desire to minimize the subjectivity of the visual judgments, color-measuring instruments can supplement and quantify visual inspection. Scientists have shown that increased proper use of instruments will resolve some of the problems of color measurement and comparison.

An additional benefit is obtained. Color measurement serves as a basis for "color memory." The unreliability of the human memory, insofar as color is concerned, is well recognized (see, for example, Bartleson, 1960). The proper use of instruments gives the manufacturer of colored materials a record of production, variation, and limits of acceptability, independent of the memory of observers.

Color *tolerances*—that is, the allowable deviations from a standard—can be established by comparing appropriate samples with a standard. In addition to the examples cited previously, where good color matches are desired, color tolerances are necessary for useful in-plant inspection of plastic materials that are made into the interior hard trim plastic parts of an automobile such as parts of an instrument panel and interior trim on doors. Current practice usually involves a visual inspection of the received goods against a reference standard prepared by those responsible for qualifying suppliers. The purpose of the in-plant inspection is often two-fold: to make an accept/reject decision and to sort the acceptable material into shade differences according to previously established specifications. Since visual judgments can be highly subjective, numerical color tolerances that supplement and quantify visual inspection are desired.

We will concentrate on the use of observers (visual judgments) and numerical color-measuring instruments in the comparison of colored samples of hard trim plastic (polypropylene) parts with those of a standard.

The statistical determination of numerical color tolerance regions useful for the above task consists of two parts: the preparation of color samples, which form a basis for the color tolerances, and the actual construction of the color tolerances.

PROCEDURE AND SAMPLE PREPARATION

For each color and material, we will assume that a reference standard has been prepared. When colored hard trim plastic parts are produced, they should agree closely enough in color to the standard to be considered an acceptable color match. The standard should be made of the same material and produced in the same way as the samples.

A visual judgment of color will be made and a numerical color-difference reading from the standard will be obtained for each sample piece. The instrument used to obtain numerical readings for the plastics is a spectrophotometer, a special type of color-measuring instrument. The readings consist of three numbers: measures of lightness, redness-greenness, and yellowness-blueness, respectively. Figure 1 shows how these numbers may be plotted on a three-dimensional graph. A sample whose numerical reading falls at the origin (the center of the graph) would have exactly the same reading as the standard and hence there should be an identical match between the two. A sample with a numerical color reading (as illustrated) whose red number is too large, and whose lightness and yellowness-blueness match, is redder than the standard. The size of this number would determine if the sample is acceptable compared to the standard. In other words, if the sample has a numerical reading that is only a little too large it may still be acceptable visually.

To establish numerical color tolerances for a "new" color (that is, one with no history), we must prepare samples that are off-color in comparison with the standard. Scientists have shown that people's reactions to various colors are not uniform. For example, for certain colors, a sample could have a red number larger than standard but still be visually acceptable. But another sample,

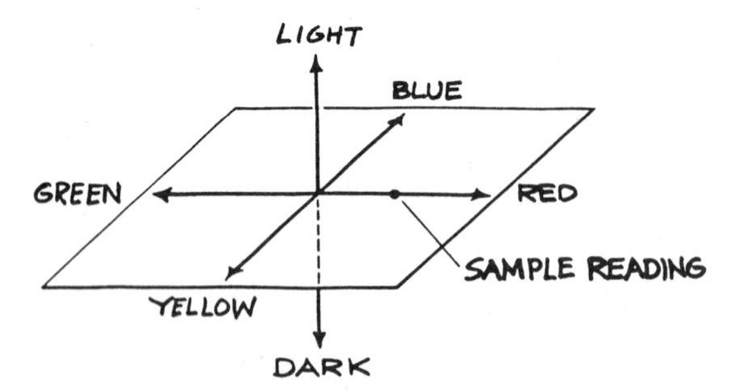

Figure 1 *Color differences in the color coordinate system. Illustrated is a sample reading that is redder than the standard but doesn't vary numerically from the standard in lightness, yellowness, or blueness.*

with a green number of the same magnitude as the red number, may be completely unacceptable visually. Also, samples that have large numerical color readings will be totally unacceptable visually. In our samples, we want some with color numbers that are unanimously acceptable visually with the standard, others that are unanimously unacceptable visually, and still others where there is considerable disagreement among the observers when they visually evaluate them against the standard. This last group of samples is extremely important because it provides us with the information to "fine tune" the boundary between the regions of acceptable and unacceptable sample numbers. As mentioned previously, the size of color numbers for acceptable and unacceptable samples differs depending on the particular color under study.

The samples can now be visually evaluated against the standard by each of several trained observers under controlled lighting conditions. The observer's task is to classify each sample as matching or not matching the standard.

As a result, for each sample, a series of visual judgments based on the number of observers and a number of independent repetitions per observer is obtained. The visual acceptability data for each sample are pooled to obtain a final decision for each sample. For example, if the majority of the observers' judgments are acceptable, the sample is accepted; otherwise, it is rejected.

The final problem in numerical color measurement is the evaluation of the color difference in terms of the acceptability of the sample as compared with the standard. Not all people agree on what the size of an acceptable difference should be. Both individual differences in color perception and personal tastes undoubtedly become important here.

Billmeyer and Saltzman (1981) state that the best procedure when customer preference appears to be playing a part is to make color measurements over a sufficiently long period that an historical record is available. If customers are at all consistent in the way they accept or reject material, it is possible to reach agreement on an acceptable color difference, even if it bears little relation to what seem to be perceptible differences.

STATISTICAL DECISION RULE

The goal is to provide an objective statistical decision rule that will characterize numerical color-difference measurements based on the visual evaluations and the numerical readings. At this point we might consider our visual judgments for each sample as the history that relates our visual evaluations to those obtained with an instrument. The method used to construct numerical color tolerances is called *discriminant analysis.* (See the Howells essay using a discriminant function.) It correlates the visual judgments and instrumental measurements of color difference and establishes numerical color tolerances around each color standard (Figure 2).

As mentioned, there are three dimensions of color. Values along the vertical axis are perceived as lightness. A second dimension of color is a measure of redness-greenness; the third dimension is a measure of yellowness-blueness.

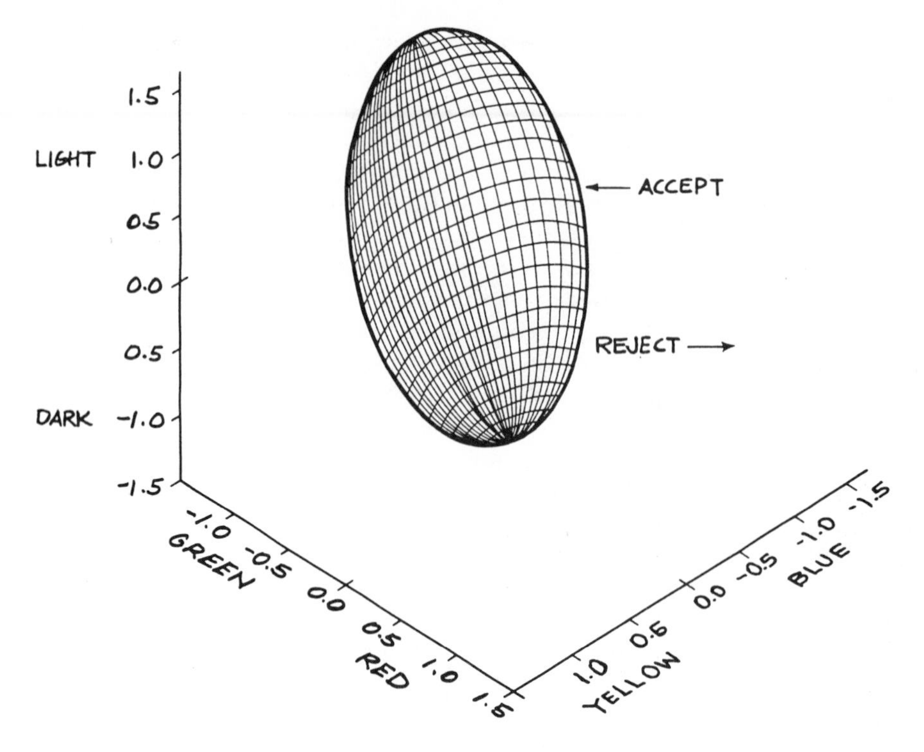

Figure 2 *Numerical color-tolerance region for dark-carmine plastic. If the point falls inside the ellipsoid, the sample is acceptable by the instrument; if it falls outside, it is rejected.*

In Figure 2 the numerical color tolerance is in the form of an ellipsoid. The visual acceptability of the deviation of the color of the sample from that of the standard depends on, among other things, the magnitude and direction of the three numbers, which are calculated with the spectrophotometer. (If the color space were uniform in all directions, then instead of an ellipsoid, the numerical color-tolerance region would be a sphere.) For all future samples of this color, we simply take an instrumental measurement of the sample and see if the numbers observed are small enough. If the reading is small enough to fall within the ellipsoid, the sample is accepted; otherwise, it is rejected.

The numerical reading from the spectrophotometer gives three numbers. For the present, consider these three numbers as three separate measurements. The rationale behind discriminant analysis is to develop a discriminant function (that is, a statistical decision rule) that will best discriminate between two groups (acceptable and unacceptable).

The discriminant function takes the three measurements from a sample, multiplies each measurement (or product of measurements) by an optimum weight specific to that measurement (or product of measurements), and then adds these products to give the discriminant score. The computation develops the optimum set of weights possible from the measurements used. By optimum weights we mean that the distinction between the visually acceptable and unacceptable sets are searched out mathematically so that the discriminant scores of the two groups are segregated one from the other to the maximum degree possible.

Although finding the discriminant function requires a great deal of mathematics, a computer makes it an easy task. Once the weights are found, it is relatively easy to obtain the discriminant scores.

EXAMPLE

To illustrate the procedure, 111 samples of dark carmine polypropylene (plastic) plaques were prepared and a corresponding standard of the same material was available. We used 16 observers to visually evaluate each of the samples against the standard. Of the 16 observers, 5 had three repetitions per sample. Consequently, each sample was given a total of 26 visual assessments. For each sample, then, there were 26 accept/reject judgments. We used a majority rule criterion: If the majority (13 or more) of the 26 judgments were accept decisions, the sample was considered visually acceptable; otherwise, the sample was judged visually unacceptable. Of the 111 samples, 40 were visually acceptable and 71 were visually unacceptable.

The numerical values for the 111 dark carmine polypropylene samples were scattered at various distances and directions from the origin (the standard). We next performed a discriminant analysis with the 111 samples to obtain a statistical decision rule, whose equation forms an ellipsoid. For any other dark carmine polypropylene sample, if its numerical reading falls inside the ellipsoid, the sample is accepted. If the numerical reading falls outside the ellipsoid, the sample is rejected.

Table 1 contains the summary results for the observers (visual decisions) and the statistical method for the 111 dark carmine samples. As stated previously, there were 40 visually acceptable and 71 visually unacceptable (reject) samples. The statistical decision rule indicates that 56 samples were acceptable and 55 were unacceptable. That is, there is disagreement between the discriminant analysis method and visual judgment in the classification of 26 samples. This translates to a 77% agreement.

REMARKS

A closer look at Table 1 indicates that very few samples rejected by the statistical method would be accepted by the human. But there is about a 40% chance that an acceptable statistical method sample will be judged unacceptable by the human. Thus this technique would also be good for prescreening, with the ultimate judgment being done by humans on about half the material not already rejected. This would cut the screening task by a factor of two.

The percentage of agreement between statistical decision rules and visual judgments for five other colors of polypropylene ranged from 88% to 96%.

Table 1 Decision table

	Statistical Method Decision		
	Accept	Reject	Total
Visual Method Decision			
Accept	35	5	40
Reject	21	50	71
Total	56	55	111

Although these percentages are considerably higher than the percentage of agreement for the dark carmine plastic, the number of samples evaluated for the five other colors ranged from 23 to 40, considerably fewer than the 111 dark carmine samples. Consequently, these percentages may be less reliable.

This technique has been used successfully to evaluate incoming shipments of colored polypropylene against a standard in automotive trim plants. If used consistently, such procedures also establish a historical record of variation and limits of acceptability of colored materials that is valuable to both the vendors and the plants.

As we have already mentioned, many industries need to match colors. Color-measuring instruments can supplement and quantify visual inspection. Such instruments determine the exact size and direction of a difference between sample and standard. Then the statistical method uses the numerical measurements to discriminate between acceptable and unacceptable samples and minimizes the subjectivity of visual judgments.

The importance of color matching cannot be overemphasized. Poor color matches are perceived by customers as poor quality. Thus acceptable color matching must be achieved in order to satisfy the customer.

PROBLEMS

1. In this study, several trained observers visually evaluated each sample against the standard and classified each sample as matching or not matching the standard. Are several visual observations of each sample necessary to perform the discriminant analysis? What is an advantage of having multiple observers?
2. Why are color-measuring instruments of value in color matching?
3. In Figure 2, the numerical color tolerance is in the form of an ellipsoid. Are there situations where the form is other than an ellipsoid? Explain.
4. What is the problem with matching color from sample to sample for acceptance of material, without reference to the original standard?
5. Each observer made a number of independent judgments of the acceptability of the samples as matches to the standard. Why is this important?

6. Give a detailed example of your own in which a discriminant function analysis might be appropriate. (Hint: See W. W. Howells's essay.)

REFERENCES

T. W. Anderson. 1984. *An Introduction to Multivariate Statistical Analysis,* 2nd ed. New York: Wiley.

C. J. Bartleson. 1960. "Memory Colors of Familiar Objects." *Journal of the Optical Society of America* 50: 73–77.

F. W. Billmeyer, Jr., and M. Saltzman. 1981. *Principles of Color Technology,* 2nd ed. New York: Wiley.

L. C. Vance. 1983. "Statistical Determination of Numerical Color Tolerances." *Modern Paint and Coatings* 73(11): 49–51.

Making Essay Test Scores Fairer With Statistics

Henry I. Braun *Educational Testing Service*
Howard Wainer *Educational Testing Service*

INTRODUCTION

As the graded tests were handed back, a crescendo of groans echoed through the classroom. After the initial shock was registered, the long-suffering teacher smiled benignly and stated, "Your poor performance, relative to previous classes, indicates that this form of the test was more difficult than I had anticipated. I'll have to curve the scores." The students' relief was palpable.

This sort of scene is common. "Curving the scores" is the transformation of the usual rules of correspondence between percent correct and its associated letter grade. In classroom tests the effect of curving almost always allows a score to qualify for a higher grade than would ordinarily be expected. While almost everyone knows this, the question of why teachers grade on a curve is shrouded in mystery. The answer, in its simplest terms, is that we curve (adjust) test scores to allow fairer comparisons among individuals who take different forms of the test.

A similar problem of adjusting test scores for fairness occurs in the subjective scoring of essays. When a large collection of essays is to be graded, it is common to engage a number of individuals to carry out the scoring, with a different sample of essays assigned to each reader. The difficulty of an essay question involves both the inherent difficulty of the question (for example: "Describe your activities over the Christmas holidays" versus "Compare Kant's metaphysics with Aristotle's") and the strictness of the reader who scores it. We can control differences of the first kind by asking everyone the same questions, but practical considerations prevent us from using the same control for the readers. Yet, if one reader has more stringent criteria than the others, those examinees who were unfortunate enough to have their exams assigned to this reader (analogous to being assigned a more difficult test form) are at a disadvantage. Fairness requires that these differences be removed (transforming/curving the ratings of the readers so that they are comparable). Readers' criteria may also shift through time; they might be more lenient on Monday than on Friday. If such variability exists, fairness requires that these day-to-day differences also be removed.

A desirable goal, then, is to develop a methodology for scoring essays so that the final grades are less affected by when or by whom each essay was read. It seems sensible to derive such grades by somehow adjusting the ratings originally given by each reader. The rest of this essay describes one solution that relies on statistical adjustment. The solution is described in the context of a testing program that includes an important essay component, the College Board's Advance Placement (AP) Program.

THE ADVANCED PLACEMENT PROGRAM

The AP Program offers specialized curricula in a wide variety of subjects, including English, American history, European history, mathematics, biology, chemistry, French, and German. High school students who participate in the program and who do well in the final examination are eligible to receive college credit for their work. In each subject, the same final examination is given all across the United States on a particular Saturday in May. Each examination has a section of multiple-choice questions and a section of free-response questions. In mathematics or chemistry, free-response questions require the student to work out solutions to problems, while in English or American history they require the student to write essays.

The answers to the free-response questions must be scored by human raters since computer programs are not yet intelligent enough to read students' handwriting and to assign values to the material. Because tens of thousands of students may write an essay on a given topic, the grading process involves bringing together as many as a hundred readers to grade papers continuously for four or five days. The readers include both high school teachers of AP courses and college teachers of those subjects. Each essay (or problem) is read by only one reader, chosen at random from the pool of available readers. He or she assigns a grade that becomes part of the total score.

PROBLEMS WITH SCORING ESSAYS

The question of whether a student is unfairly advantaged (or disadvantaged) by having his or her essay read by one particular reader rather than another, is a critical issue. Readers, being human, will differ in their judgments of the quality of a particular essay and so the score assigned to that essay will depend to some extent on the "luck of the draw." This dependence on chance is undesirable and should be eliminated to the extent feasible.

Before we can act to eliminate this variability we have to understand how it can arise. First, different readers may have different scales for scoring. That is, two readers may agree on how to rank a set of papers but one might systematically assign higher grades than the other. Second, two readers may assign the same scores on average but generally disagree on which essays deserve high grades and which low. In practice, both kinds of discrepancies, as well as others, will occur to some extent.

Because the grading process extends over a number of days, the score assigned to an essay may also depend on when it is graded. There may be, for example, a general trend to grade more leniently (or more stringently) over the course of the week. Beyond this general trend, individual readers will exhibit their own trends through time. Such global patterns in assigned scores have nothing to do with the quality of the essays. If these patterns exist, they also contribute to the role that chance plays in the grade assigned to a student's essay.

Nonstatistical provisions are currently in place to minimize the potential impact of these factors on the grades. The AP Program carefully trains readers before the scoring sessions begin and continuously monitors them during the sessions. For each subject, a chief reader with several years experience in the program is appointed to take responsibility for the integrity of the scoring process. Soon after the answer booklets are returned, the chief reader selects a number of essays to illustrate different levels of the score scale. After extensive discussions with the senior readers and, eventually, with all the readers in the pool, the chief reader constructs a detailed list of criteria. Adherence to this "rubric" is monitored by periodically asking all the readers to grade the same paper. If substantial discrepancies occur, the readers undergo further training. This approach seems to work reasonably well but, as we shall see, there remains room for improvement.

Before we go on to discuss how statistical thinking can help, we must have some way of measuring how well a suggested approach succeeds in reducing the role of chance in grade assignment. This will provide us with a yardstick by which to judge the effectiveness of a new method.

HOW WELL ARE WE DOING?

Unfortunately there is no simple way of getting a "true score" for an essay, so we cannot simply compare the assigned score with "truth" and use the difference as an indication of the influence of chance. If an essay were read by all the readers in the pool, then the average of these scores could be used in place of a true score. It would be impractical, however, to obtain so many readings except for a very few essays.

Scientists who study test scores have followed a rather different strategy. They judge the merit of a scoring procedure by applying it twice to a large sample of essays and assessing the agreement between the two sets of results. In the case of AP, they might select a sample of 500 essays for the "experiment." Each essay would then be scored twice—each time on a day and by a reader chosen at random. Using the first set of scores, the essays would be listed from high to low. A second ranking would be obtained from the second set of scores. If the role of chance is relatively small, then an essay should fall at about the same place in the two lists. But if chance makes a large contribution, then the two rankings will differ considerably.

The level of agreement between rankings is usually measured by a quantity called the ***reliability coefficient.*** The reliability coefficient is a number that is calculated from the numerical information contained in the two lists. In this setting, it can range from near 0 to near 1. If there is little agreement between the two lists, the coefficient will take on a value near 0, indicating that chance is playing a substantial role in the grading process. On the other hand, if there is substantial agreement between the lists, the coefficient will take on a value near 1, indicating that chance is playing a minor role.

Typical values of the reliability coefficient for essay scores in the humanities are between .3 and .6. For problems in chemistry, the reliability coefficient usually lies between .6 and .8. To get some feeling for what these numbers mean, consider the following findings. If a group of boys are ranked by height at age six and then again at age ten, the reliability coefficient for the two lists is greater than .8. If the boys are ranked by performance on an objectively scored intelligence test at two different ages, the reliability coefficient is usually greater than .6. Finally, if boys of the *same* age are ranked once by height and again by performance on an intelligence test, the reliability coefficient for the two relatively unrelated lists is usually only about .2 or .3.

It is not unusual (such as in the study described below) to have more than just two rankings of a set of essays. In situations like this we can reduce the many rankings to just two by simply choosing any two at random and calculating the reliability as before. Later, when we talk of the reliability of a particular scoring procedure, we will be referring to a measure that is closely akin to a pairwise reliability averaged over all pairs of judges.

CALIBRATING READERS AND DAYS

We are now ready to see how statistical thinking can help. The basic idea is to reduce the effect on scoring reliability of some of the sources of variability we have mentioned: systematic differences between readers or between days. By that we mean the following: if we knew, for the same set of papers, that one reader would assign scores that were on average 10 points higher than another reader's, we could adjust the first set of scores by subtracting 10 points from each of them (or by adding 10 points to each of the scores in the second set). The two sets of scores would thus have the same average. This is as it should be since they refer to the same set of papers.

Exactly the same sort of adjustments could be used to deal with systematic differences between days. If a set of papers graded on one day received scores that were on average, say, 5 points lower than they would receive on another day, we could add 5 points to the first set of scores to make the averages equal. The process of making averages equal is called *calibration.* In the context of essay scoring, calibrating both readers and days would improve the reliability of the scores by eliminating two sources of chance variation. The degree of improvement would depend on, among other things, how large these differences were in the first place.

COLLECTING DATA

Where are we to obtain the information that we need to carry out the calibration? In the operational grading, each paper is only read once—by a particular reader on a particular day. If readers assign different average grades over the course of the five-day grading period, we do not know whether to attribute those differences to real and consistent differences among the readers, or to differences in the quality of the essays they happened to read, or both. To make some progress, we will have to collect specialized data that will give us the information we need.

Statistical theory can guide us to the design of an experiment that will efficiently collect those data and tell us how to use them appropriately. Consider the following experiment. Suppose we choose a small sample of essays at random from among the pool of tens of thousands available and arrange to have each essay read by each reader on each day. The data thus obtained would allow us to estimate average differences among readers as well as average differences among days. (We use the term estimate because we would have observed the grading behavior of the readers only for the sample and not for all the essays.)

We could use these estimates, obtained from this small sample of essays, to calibrate readers and days. That means we could adjust the scores for the entire pool of essays, by whomever and whenever they were graded, based on the information collected in the experiment. But before we do that we have to consider carefully the quality of the information we would be using.

This experiment presents at least two problems. Because of the enormous number of readings that have to be carried out, there is a severe restriction on the number of extra readings that can be added for the experiment. Since each reader is to read each essay on each day, the number of essays has to be kept very small—say, five to ten. This raises questions about the representativeness of the results: Would we get substantially different estimates if we chose another set of five essays? A second issue arises from the repeated readings of the essays. To the extent that readers remember the score they assigned to an essay on the previous day and just copy it, we are not collecting bona fide information. Such distortions in the estimates could result in our making adjustments in the wrong direction, so that calibrations would lower reliability rather than raise it!

STATISTICS TO THE RESCUE

Our aim is to estimate the relative stringency of the different readers as well as the scoring trends across time without encountering the pitfalls mentioned above. Fortunately statisticians have devoted a lot of effort to solving problems of this sort. They have developed special methods for efficiently collecting data called ***experimental designs.*** An example of a design that meets our needs is contained in Table 1. The table represents a set of instructions for allocating readings for a four-day experiment involving 12 readers and 32 essays chosen at random from the pool. (One of the reasons that the numbers 12 and 32 were chosen is that they are both divisible by 4, the length of this particular experiment; other combinations are possible.) Each of the 32 rows corresponds to

Table 1 Allocation plan of essays to readers

	Readers											
Essays	*1*	*2*	*3*	*4*	*5*	*6*	*7*	*8*	*9*	*10*	*11*	*12*
1	1*	1	1	4	4	4	3	3	3	2	2	2
2	3	3	4	2	2	3	1	1	2	4	4	1
3	4	2	3	3	1	2	2	4	1	1	3	4
4	2	4	2	1	3	1	4	2	4	3	1	3
5	1	4	4	4	3	3	3	2	2	2	1	1
6	3	2	1	2	1	4	1	4	3	4	3	2
7	4	3	2	3	2	1	2	1	4	1	4	3
8	2	1	3	1	4	2	4	3	1	3	2	4
9	1	2	2	4	1	1	3	4	4	2	3	3
10	2	3	1	1	2	4	4	1	3	3	4	2
11	3	4	3	2	3	2	1	2	1	4	1	4
12	4	1	4	3	4	3	2	3	2	1	2	1
13	1	3	3	4	2	2	3	1	1	2	4	4
14	3	1	2	2	4	1	1	3	4	4	2	3
15	4	4	1	3	3	4	2	2	3	1	1	2
16	2	2	4	1	1	3	4	4	2	3	3	1
17†	1	1	1	4	4	4	3	3	3	2	2	2
18	3	3	4	2	2	3	1	1	2	4	4	1
19	4	2	3	3	1	2	2	4	1	1	3	4
20	2	4	2	1	3	1	4	2	4	3	1	3
21	1	4	4	4	3	3	3	2	2	2	1	1
22	3	2	1	2	1	4	1	4	3	4	3	2
23	4	3	2	3	2	1	2	1	4	1	4	3
24	2	1	3	1	4	2	4	3	1	3	2	4
25	1	2	2	4	1	1	3	4	4	2	3	3
26	2	3	1	1	2	4	4	1	3	3	4	2
27	3	4	3	2	3	2	1	2	1	4	1	4
28	4	1	4	3	4	3	2	3	2	1	2	1
29	1	3	3	4	2	2	3	1	1	2	4	4
30	3	1	2	2	4	1	1	3	4	4	2	3
31	4	4	1	3	3	4	2	2	3	1	1	2
32	2	2	4	1	1	3	4	4	2	3	3	1

*The entries in the table indicate the day that reader scored that essay.
†Rows 17–32 are just duplicates of rows 1–16.

a different essay, while each of the 12 columns corresponds to a different reader. The numbers in each row of the table indicate which readers are assigned to read that essay on that day. For example, reader 1 grades essay 16 on day 2.

This design calls for each of the 32 essays to be scored three times each day, for 96 readings altogether. (Note that if each reader were required to score every essay each day within an overall limit of 96 readings, only 8 essays could be included in the experiment. By relaxing this requirement, we are able to employ four times as many essays.) The allocation of readers to essays is not done in a haphazard way. In fact, there is a delicate choice of reader-essay combinations that enables us to obtain estimates of systematic differences among readers, even though no two readers read exactly the same set of papers.

Over the course of this four-day experiment, each reader will read each of the 32 essays exactly once. Consequently, there are no repeat reading, or carry-over, effects to worry about. Since each essay is read three times each day and each reader reads eight essays each day, we can also obtain estimates of systematic differences between days. Because our estimates are based on a sample of 32 essays, rather than the eight essays that would be the limit with a complete design involving the 96 readings, they should be more representative as well. With the particular design we have chosen, it is even possible to make useful comparisons between readers on a day-by-day basis.

SOME RESULTS

To get a flavor of the results, we present the findings of one such experiment carried out for an essay question in English Literature and Composition for which scores were on a scale of 100 (low) to 900 (high). Table 2 shows for each day the average scores assigned to essays graded on that day as well as the differences between these day averages and the overall average for the entire experiment. On day 1, for example, the average score was 490, which is 7 points higher than the overall average of 483.

Ideally the day averages should be very similar and indeed they are in this case. (The largest difference among days is 12 points. This is less than 3% of the average score in the experiment.) But this means that there is very little to be gained in trying to adjust for systematic differences among days—there just aren't any!

Table 2 Daily averages and their deviations from the mean

Day	*Day Average*	*Day Average Minus Experiment Average*
1	490	7
2	479	−4
3	478	−5
4	485	2
Experiment Average	483	

On the other hand, Table 3 presents the average score assigned by each reader over the course of the experiment. To see the substantial differences more clearly, we also show the differences between these reader averages and the overall average of 483. Reader A, the most lenient reader, typically scored essays 82 points higher than the average while reader L, the most stringent reader, typically scored essays 58 points lower than the average. Remember these are just estimates of the differences in scoring levels between readers based on 32 readings. Nonetheless, they have considerable credibility because through our design we have been able to balance out sources of variation that could otherwise degrade the estimates.

It certainly appears as if, for this question at least, days don't matter much but readers do. We have to remember, though, that there are three times as many scores contributing to a day average as there are contributing to a reader average. Accordingly, some proportion of the greater variability we observe among reader averages (as compared to day averages) may be due to the vagaries of chance. However, we can capitalize on the features of this particular design and the methods of statistical hypothesis testing to properly compare the relative variation of readers versus days. When we do, we find that our first, naive impressions are justified: readers matter much more than days.

To carry out the calibration, then, we subtract 82 from all the essays graded by A, subtract 61 points from all the essays graded by B, and so on. We can judge the effectiveness of the procedure by comparing the reliability of the original scores with that of the adjusted scores. The former is .57 and the latter is .61, a difference of .04. That doesn't sound like a great improvement for all that effort. The following calculations may help put the gain in some perspective.

By using some mathematical analysis it is possible to show that if each essay had been read independently by two readers and the average of the two scores

Table 3 Reader averages and their deviations from the mean

Reader	*Reader Average*	*Reader Average Minus Experiment Average*
A	565	82
B	544	61
C	517	34
D	506	23
E	487	4
F	484	1
G	476	−7
H	473	−10
I	454	−29
J	432	−51
K	432	−51
L	425	−58
Experiment Average	483	

To ease comprehension of this table, readers have been ordered by the average grade that they assigned.

used as the final score, then the reliability of these averaged scores would be about .73. (Obtaining multiple readings is the standard way of improving reliability.) Our gain of .04 is 25% of .16 = .73 − .57, the gain in reliability possible with double reading.

Remember that with the information gleaned from this little experiment we can adjust the scores of the entire collection of essays submitted. Our data have been obtained at a small fraction of the cost of hiring enough extra readers to double read the tens of thousands of essays on hand. We estimate the cost factor will typically be about one-thirtieth. Since we have achieved one-quarter of the gain at one-thirtieth the cost, a cost/benefit analysis would yield a factor of seven or eight in favor of the calibration approach. This means that if it cost, say, $5,000 to run the experiment, it would have cost about $150,000 to hire enough readers for a complete double reading, and so one-quarter of that amount ($37,000) would be required to achieve the same gain in reliability. This suggests that using calibration should be seriously considered.

SHOULD CALIBRATION BE USED?

Calibration experiments have now been carried out on five different AP examinations. In general, calibrated scores exhibit enhanced reliability—especially when the reliability of the original scores is on the low side to begin with. In one case the estimated reliability of the calibrated scores actually exceeded the projected reliability of double reading! The obvious success of such an experiment, however, is not sufficient to guarantee the operational implementation of the procedure. There are many other issues to be addressed.

One such issue arises because the experiment we have described requires considerable planning and analysis. We have also investigated another calibration procedure that can better be adapted to the tight time schedules that must be met in reporting scores to candidates. It is one we previously mentioned; namely, to make the adjustments on the basis of the operational grades. For example, if the average grade assigned by a reader over the entire grading period was 10 points higher than the average grade for all readers, we would then subtract 10 points from all the scores that grader assigned. A potential difficulty with this approach is that this reader, by chance, may have been assigned essays that were typically better than average and deserved the higher scores. In that case, an adjustment of 10 points would be too large.

In practice, the essay booklets undergo various stages of haphazard shuffling before landing on a reader's table. Unfortunately we have no direct way of determining whether readers typically receive representative (truly random) samples of essays. But this is precisely where our experiment plays an important role. We can compare the calibration using the operational scores and the calibration using the results of the experiment (in which the sample of essays is controlled and the randomization is carefully executed). When we do, we find the results are very much the same. This gives us confidence in the simpler, cheaper method.

It is worth pointing out that the data collected in an experiment such as the one we have described can lead to insights that are just not available from the operational data. Using methods of analysis that are too technical to be described here, we can learn more about the relative contributions of readers and days to score unreliability—and do it in a way that facilitates comparisons across different tests. We can also estimate the upper limit of reliability that can be achieved through calibration. This gives us a meaningful target to shoot for.

In addition to considerations of feasibility, we also have to take into account the possible reactions of both students and schools to the notion of statistical adjustment of scores. Since the first phase of this research has clearly established that a statistically designed experiment can make the process of grading essays more fair, it only remains to iron out these other aspects before adopting its use widely. As this essay is being written (December 1987) this decision process is under way and may be operationally in place as you read about it.

PROBLEMS

1. Does training of essay raters yield the result that all readers will score the same essay identically? Why or why not?
2. Would you expect essay readers to change their scoring scale over the course of the week?
3. Why is calibration of essay readers necessary?
4. Why can't we just have all essays read by several readers?
5. What is the advantage of using the complex experimental design in Table 1 rather than just having all experimental essays read by each of the readers on each of the days?
6. How much accuracy is gained by adjusting for differences in reader performance?

REFERENCES

H. I. Braun. 1988. "Understanding Score Reliability: Experiments in Calibrating Essay Readers." *Journal of Educational Statistics* 13:1–18. This contains a full description of the experiments and procedures summarized in this essay.

W. G. Cochran and G. M. Cox. 1957. *Experimental Designs,* 2nd ed. New York: Wiley, Chapter 9. A classical work on experimental design.

H. O. Gulliksen. 1987. *Theory of Mental Tests.* Hillsdale, N.J.: Erlbaum. (Originally published in 1950 by John Wiley & Sons.) This was the first (and perhaps still the most readable) comprehensive account of mental test theory—see especially pages 211–214 for a description of how to grade essays.

Statistics, Sports, and Some Other Things

Robert Hooke *Westinghouse Research Laboratories*

Spectator sports provide more than just observation of athletes who perform with admirable skill. There is, for example, the drama of a young quarterback trying to lead a professional football team for the first time in front of 70,000 onlookers or that of a veteran pitcher calling on his experience to augment his dwindling physical resources in a crucial game of a close pennant race. Because these dramas are truly "live" and unpredictable, they are much more fascinating to some people than the well-rehearsed performances of the stage.

Not every moment in sports is dramatic, of course, but throughout any contest between professionals, the spectator is privileged to watch a group of people carrying out their jobs almost in full public view, to see how they meet their problems and how they react to their own successes and failures. Baseball and football provide especially good opportunities for such observation because each of these games consists of a sequence of plays, as opposed to the fairly continuous action of basketball, hockey, soccer, and racing sports. Spectators see more than strikeouts and home runs, completed passes and interceptions. They see a manager gambling on a hit-and-run play or a quarterback deciding to pass for the first down he desperately needs. Fans have opinions on what their team should do in various situations, and they watch to decide whether their manager or coach is a good strategist or a poor one.

Management in professional sports has many similarities to management in business and industry. Some managers and coaches are smarter than others, and some make use of more advanced methods than others do. This is true in sports, as it is in business, in spite of the folk wisdom of the sports pages that often maintains that all managers and coaches are pros and about equally good. Some can get a great deal out of inferior personnel, but none can overcome more than a certain amount of incompetence among the people who work for them. Some are natural gamblers, some are always conservative, and only a few are intelligent enough to be one or the other depending on what the circumstances call for. Sports managers are different from business managers mainly in that their actions are so much more visible. Because of this visibility we should all be able to learn by watching them as they make their various moves in the goldfish bowl of professional sports.

STATISTICS AND MANAGEMENT

What does all this have to do with statistics? The real concern of statistics is to obtain usable quantitative information, especially about complex situations that involve many variables and uncertainties. "Usable" means that its purpose is to help us to improve our behavior in the future, that is, to help us learn how to extract from these situations more of whatever it is that we're trying to get. Some managers make good use of statistics and some don't; this is true whether they manage factories or baseball teams.

Suppose, for example, that we are manufacturing rubber tires. An expert will no doubt be able to detect from the example that I know nothing about the tire business; in fact, I chose this example because I've never been in the tire business and hence will not implicate real people. At some point in the process, let's suppose that we have a mass of liquid rubber that will ultimately be turned into tires. Being aware that this batch may possibly have been improperly prepared, we would like to test it in some way so that, if it is defective, we can throw it away without wasting money processing it into defective tires. Unfortunately, the true test of a tire is a road test, and we can't road test a batch of liquid, so we must perform some test that we think is relevant, such as a viscosity measurement. This measurement will take time and cost money, and sooner or later someone will raise the question "Is it worth the money we're spending on it?" This is always a good question, and it usually leads to much heated debate. The debate will include arguments based on intuition, experience, laboratory tests, and scientific theory; each has its place in the process of seeking after the truth, but they are basically predictors, and the only way to be sure of what will happen in the field is to see what actually happens in the field. This means that we should collect statistics. After measuring a batch, we should follow it through the manufacturing process and see the quality of a sample of tires made from this batch. We repeat this on another batch, and so on. After a while we can establish the relationship between tire quality and viscosity, and we can use this to determine whether to continue with the test, taking into account the testing costs, the cost of the manufacturing process that follows the test, the value of a good finished tire, and so on. Now you may say, of course, that this method isn't infallible because it involves sampling and, hence, sampling error (see the essay by Neter) and because the process may change unexpectedly, and so on. But this method gets us as close to the truth as we can come, and this final objection merely says that you never have it made, even if you use statistics.

THE STRATEGY OF BUNTING

The student of management behavior can find many instances of this type of problem in sports, and such a student, if smart, can profit from the mistakes that are made visibly on the diamond or the gridiron. Take, for example, the sacrifice bunt in baseball. There are those who swear by it and there are those who seldom use it. They engage in passionate arguments as to whether it is a good strategy. As we shall see, statistics can't settle the issue once and for all, but it can shed a great deal of light on the problem, and most of the argument could be eliminated if people would look at some of the facts.

The sacrifice bunt is a play that is used to advance a base runner from first base to second, or from second to third, normally sacrificing the batter, who is thrown out at first. Many managers use the sacrifice bunt routinely, and they refer to their behavior as "percentage baseball," as if they knew the percentages, which, apparently, most of them do not. The routine is that you bunt if there is a man on first or second, nobody out, and your team is only slightly ahead, tied, or not "too far" behind. One or more runs behind is considered too far for the visiting team, and two or more runs behind too far for the home team, the difference coming from the fact that the home team bats last and can afford to "play for a tie."

Why does the manager decide to bunt? Ultimately, of course, he does it to win more games. At the moment of doing it, he is trying to increase the chance of getting one additional run while giving up some of the opportunity to get several. The theory is simple. It takes at least two singles or a double to score a runner from first, while a runner on second can usually score on a single alone. In addition, if a runner on second can be moved to third with only one out, a score will result from any hit, error, wild pitch, passed ball, long fly, balk, or slow grounder. Proponents of bunting are fond of quoting this list, but it contains some fairly rare events, and this raises the real questions—when we use the bunt, by *how much* is our chance of scoring one run increased, and *how much* do we sacrifice in terms of possible additional runs? Again, the only way to get an answer to this question that is relevant to real major league players playing under the pressure of real games is to take statistics from actual games. There is no way to provide realistic conditions for an experiment, and theory (see Cook, 1966; Hooke, 1967) is of dubious value.

Although records of games played exist in the archives of organized baseball, turning these into usable data is a major task that, if it has been done, has not been made public to my knowledge, except in Lindsey (1963). Lindsey discusses records of several hundred major league games played in 1959 and 1960, and he produces some very interesting statistics, some of which are shown in Table 1.

To see how we, as armchair managers, would use this table to decide about bunting, let's look at the first two lines. (For the moment, we'll think only of average cases, but no good statistician dwells on averages alone, so we'll discuss special situations later.) We start, say, with a man on first and no outs. The table says that this situation was observed 1,728 times (occasionally, perhaps, more than once in the same inning). In a proportion of these cases equal to 0.604, no runs were scored during the remainder of the inning; that is, in 1,044 cases no runs scored, and 1,044/1,728 = 0.604. This means also that the proportion of times at least one run scored from this situation is 1 — 0.604, or 0.396. We use these proportions as estimated probabilities of the various events; thus near the end of a tight game, the number 0.396 measures the average "value" of having the situation of a man on first and no outs. For earlier parts of the game, the value is more closely related to the number of runs that are scored in an inning, on the average, starting from this situation; this is given in the fourth column as 0.813 for the situation in question.

Now if we make a sacrifice bunt that succeeds in the normal way, the runner on first will move to second and there will be one out. Is this a better situation than we had? In the sense of average number of runs scored, it is decidedly worse; the first and second lines of Table 1 show that the average number of runs scored from the man-on-first-no-out situation is 813 per thousand, but from the man-on-second-one-out situation, it is only 671 per thousand. On the average, then, a normally successful bunt loses 142 runs per 1,000 times it is tried. But what about the last inning of a tight game when we only care what has happened to the probability that at least one run will be scored? This figure

Table 1 Relation of runs scored to base(s) occupied and number of outs

Base(s) Occupied	*Number of Outs*	*Proportion of Cases No Runs Scored in Inning*	*Proportion of Cases of at Least One Run Scored in Inning*	*Average Number of Runs in Inning*	*Number of Cases*
1st	0	0.604	0.396	0.813	1,728
2nd	1	0.610	0.390	0.671	657
2nd	0	0.381	0.619	1.194	294
3rd	1	0.307	0.693	0.980	202
1st, 2nd	0	0.395	0.605	1.471	367
2nd, 3rd	1	0.27	0.73	1.56	176

Source: Lindsey (1963).

has dropped from 0.396 to 0.390; these numbers are so close that their difference is readily explained by chance fluctuation from the sample. So we conclude that the advantage of having the runner go from first to second is almost exactly canceled by the disadvantage of the additional out that typically occurs on a bunt play.

Conclusion. On the average, bunting with a man on first loses a lot of runs. On the average, it doesn't increase the probability of scoring at least one run in the inning. Here we've assumed that the batter is always out at first, but, of course, he is sometimes safe, thereby increasing the efficacy of bunting. It is probably more often true, however, that the front runner is thrown out at second, a disaster to the team that chose to bunt. It would appear that bunting with a man on first early in the game should be done only when it so takes the defense by surprise that the chance of the batter's being safe is substantial. Even late in a tight game there is no visible advantage to such bunting unless special circumstances prevail.

Now let's think of the problem of the man on second with nobody out. The table tells us that he will score (or at least somebody will score) in all but 381 cases out of 1,000, that is, in 619 cases out of 1,000. *If* we can move him to third by sacrificing the batter, we can raise the 619 to 693. (Note that we lose 214 runs per 1,000 tries doing this, but let's again consider the case where it is late in the game and we need only one run.) Here there is indeed something to be gained by a successful bunt play, but it's time to face reality: the bunt play doesn't always work. How often it works depends on a lot of things, and we don't have statistics for an average result, but let's see how we would use them if we did.

If the batter bunts the ball a little too hard, the defending team happily fires the ball to third base and the lead runner is put out, leaving the offensive team with a man on first and one out, their probability of scoring at least one run having gone from 0.619 to 0.266, the latter figure coming from the complete table in Lindsey's paper. The typical manager does not admit the possibility of such an event. After it happens he dismisses it with the remark "These young fellows don't know how to bunt like we used to." I know this remark was being made before any of today's managers were making their first appearance as professional players, and it probably originated in the nineteenth century by the first nonplaying manager. The remark is merely an excuse for not studying the problem, but let's not be too hard on baseball managers; we have pointed out already that the moves they make in plain sight are duplicated by other kinds of supervisors in less visible circumstances.

As I said above, we don't have statistics for the results of a bunt try with a man on second, so I'll make up some, trying to be as realistic as possible from unrecorded personal observations over the years. Here they are:

1. 65% of the time the runner moves to third, and the batter is out (normal case).
2. 12% of the time the runner is put out at third, and the batter is safe at first.

3. 10% of the time the runner must stay at second, and the batter is out, for example, when the batter bunts a pop fly or strikes out.
4. 8% of the time the batter gets on first safely, that is, he gets a hit, and the runner also advances.
5. 5% of the time the bunter hits into a double play, that is, he and the runner are both thrown out.

Now to compute the overall probability of scoring at least one run, we simply multiply and add according to the rules of probability. If you don't know these rules, do it this way: start with 1,000 cases. In 650 (that is, 65%) of these we have result 1, namely, a man on third and one out. The table says that he will score 69.3% of the time, so we take 69.3% of 650, and get 450. That is, in 450 cases the bunt succeeds as in 1, and a run ultimately scores. Now in 120 cases the outcome is as in 2, and Lindsey's complete table says that a score then occurs 26.6% of the time. So we take 26.6% of 120 and get 32. Add this to the 450 and keep going. What we get for all five cases, using Lindsey's complete table where necessary, is

$$
\begin{aligned}
0.693(650) + 0.266(120) + 0.390(100) + 0.87(80) + 0.067(50) \\
= 450 + 32 + 39 + 70 + 3 \\
= 594.
\end{aligned}
$$

In other words, we will get at least one run in only 594 cases out of 1,000. Before the bunt our chances were 619 out of 1,000, so we have shot ourselves down. Of course, if our hypothetical data in 1 to 5 above are too pessimistic, the correct result will be a little more favorable to bunting, but it would appear that any realistic estimates will lead to the conclusion that bunting is not profitable on the average.

The intelligent use of statistics requires more than just a look at the averages. The above data and accompanying arguments show that bunting, used indiscriminately as many managers do, is not a winning strategy. This doesn't mean that one should never bunt, however. The man at bat may be a weak hitter who is an excellent bunter, and the man following him may be a good hitter; the batter may be a pitcher whose hitting ability is nil, but who can occasionally put down a good bunt; or the other team may clearly not be expecting a bunt, so that the element of surprise is on our side to help the bunt become a base hit. In any of these cases the bunt can be a profitable action. The role of statistics is to show us what our average behavior should be. In general, if the average result of a strategy is very good, we should use it pretty often. If the average result is poor, we should use it sparingly, that is, the special circumstances that lead to it should be very, very special. There are those who say that statistics are irrelevant and that they treat every case as a special case. This is probably impossible, and if such people would examine their behavior over a long period of time, they would probably find it quite statistically predictable. Incidentally, if one takes the point of view that surprise is the whole thing, that is, that the objective is to be unpredictable, then a randomized strategy is indicated; this is elaborated in any book on mathematical game theory.

THE STRATEGY OF THE INTENTIONAL WALK

Another strategic move in baseball is the intentional base on balls. The opposing team has a man on second, say, with one out, and we decide to put a man on first intentionally, either to try to get a double play or to have a force play available at all three bases. Is it worth it? Lindsey's table shows that before the intentional walk the probability of scoring at least one run is 0.390, but afterward it is 0.429. Clearly, on the average, the intentional walk is a losing move; followed by a double play it's great, but followed by an unintentional walk it can lead to a calamity, and the latter possibility is part of the reason for the numerical results just quoted. Widespread use of the intentional walk seems to be based on sheer optimism, as the statistics appear to show that the bad effects, from the point of view of the team in the field, definitely outweigh the good ones, on the average. What about special cases? If the batter is a good one, to be followed by a poor one, then the data don't necessarily apply, and the intentional walk may be a good thing. It probably should seldom be used early in the game, though, unless the following batter is a weak-hitting pitcher because it causes the average number of runs to go up from 671 per 1,000 to 939 per 1,000 owing to the additional base runner, and it is doubtful that there are many special cases that are so special as to outweigh this fact. At the end of the game, the data of Table 1 may be too optimistic in favor of the intentional walk for this reason: great additional pressure is placed on a pitcher who gives an intentional walk to fill the bases with the potential game-ending run on third, and this pressure seems to produce an increase in the frequency of wild pitches, hit batsmen, and unintentional walks.

COLLECTION AND USE OF DATA

Figures such as those in Table 1 are obviously of little value unless they are based on a rather large number of cases. It isn't at all obvious, though, how large the number of cases should be. Mathematical statistics answers questions about how large sample sizes should be, but the questions must be specific. We can't, as we are sometimes asked to do, say that 100 (or 1,000 or 2,000) is a good all-around sample size. If, however, we are asked to find the probability of at least one run resulting from a man on first with no outs, we can, with certain reasonable simplifying assumptions, determine how large the sample must be so that we can be 90% sure, for example, of being within 0.005 of the correct answer. Table 1 shows in the last column the sample size that was used to produce the data of the earlier columns. For an individual keeping records as a pastime, this represents a major effort. We would think that baseball people, engaged in a competition in which a few extra victories can make a difference of a great deal of money, would go to the trouble to collect even larger samples. They wouldn't want to go too far in this direction, however, because information tends to become obsolete. Changing rules, playing fields, and personnel cause the game to change slightly from year to year. Sometimes scoring is relatively low for a few years, and then it increases for a few years. Data gathered in one of these periods of time may not be altogether valid as a basis for decisions in another.

Data of the sort we have been talking about here are sometimes called *historical* as opposed to *experimental,* or *controlled.* The distinction is important in many areas. For example, if statistics are produced showing that smokers have lung cancer with much higher frequency than nonsmokers, this *historical* fact *in itself* does not demonstrate that smoking increases the lung cancer rate. (After all, children drink more milk than adults, but this is not why they are children.) The problem is that there may be other variables that, for example, help cause lung cancer and also influence people to become smokers. Nevertheless, the historical statistics on cancer were very suggestive and led to various experiments in laboratories that have strengthened most people's belief in a causal relationship. We can make good use of historical data, in other words, but we must be careful about inferring cause-and-effect relationships from them.

No doubt because of frustrations in trying to draw conclusions from historical data, statisticians developed the science and art called the *design of experiments.* If we can do a properly designed experiment, we are in a much better position to draw valid conclusions about what causes what, but the possibility of a designed experiment is not always open to us. When we can't experiment, we must do what we can with available data, but this doesn't mean that we shouldn't keep our eyes open to the faults that such data have.

CONCLUSIONS

So what have we learned from our look at sports statistics? We have learned these do's and don'ts:

1. Don't waste time arguing about the merits or demerits of something if you can gather some statistics that will answer the question realistically.
2. If you're trying to establish cause-and-effect relationships, do try to do so with a properly designed experiment.
3. If you can't have an experiment, do the best you can with whatever data you can gather, but do be very skeptical of historical data and subject them to all the logical tests you can think of.
4. Do remember that your personal experience is merely a hodgepodge of statistics, consisting of those cases that you happen to remember. Because these are necessarily small in number and because your memory may be biased toward one result or another, your experience may be far less dependable than a good set of statistics. (The bias mentioned here can come, for

instance, from the fact that people who believe in the bunt tend to remember the cases when it works, and vice versa.)

5. Do keep in mind, though, that the statistics of the kind discussed here are averages, and special cases may demand special action. This is not an excuse for following your hunches at all times, but it does mean that 100% application of what is best on the average may not be a productive strategy. The good manager has a policy, perhaps based on statistics, that takes care of most decisions. The excellent manager has learned to recognize occasional situations in which the policy needs to be varied for maximum effectiveness.

Since this article was written, computers have invaded sports just as they have many other fields of activity, and greatly increased use of statistics has followed. Consequently, some of my remarks about the use of statistics, particularly in baseball, are no longer true as stated. Also, because of changes in the game ranging from new ballparks to rule changes such as the introduction of the designated hitter, the 1959–1960 data of Table 1 may be obsolete for one of the major leagues. All this means that the examples are out-of-date while the general principles remain true. In fact, a new similarity between sports management and business management has been added, namely, that the increased use of statistics has not necessarily resulted in more intelligent use of statistics. Since my retirement in 1979 I have not followed major league baseball at all closely, but from watching postseason games I've concluded that few, if any, managers have used the increase in available knowledge to improve their strategies.

PROBLEMS

1. Refer to Table 1. In how many cases with a man on third and one out, did no runs score?

2. Suppose second base is occupied and there are either no outs or one out. In how many of such cases in Table 1 are no runs scored in the inning?

3. Suppose there are runners on first and second, no outs, and it is early in the game. Assuming the batter will be out and the runners advance one base, do the figures in Table 1 suggest a bunt? Explain your answer.

4. Suppose there are runners on first and second, no outs, and it is the last inning of a tight game. Assuming the batter will be out and the runners advance one base, do the figures in Table 1 suggest a bunt? Explain your answer.

5. Suppose the statistics for the results of a bunt try with a man on second are 70%, 13%, 9%, and 3%, respectively, instead of 65%, 12%, 10%, 8%, and 5% assumed by the author. Would bunting then be profitable on the average in this situation? Explain your answer.

6. When might a sacrifice bunt be a wise move in a situation where, on the average, it is not?

7. a. Distinguish between *historical* and *experimental* data.
   b. Why didn't Lindsey conduct a controlled experiment?

8. Use the following additional statistics from Lindsey and the outcome percentages given in the text. Assume there is a man on second and one out. The batter attempts a bunt.

Base Occupied	*No. of Outs*	*Probability That No Runs Score in the Inning*
1	2	.886
2	2	.788
3	2	.738
1, 3	1	.367

   a. How many times (out of 1,000 cases) will at least one run score?

   b. How does possibility 5 (bunter hits into double play) enter your calculation?

REFERENCES

E. Cook. 1966. *Percentage Baseball.* Cambridge, Mass.: MIT Press.

R. Hooke. 1967. Review of Cook (1966). *Journal of the American Statistical Association* 62:688–690.

R. Hooke. 1983. *How to Tell the Liars from the Statisticians.* New York: Marcel Dekker. (See especially sections 18, 25, 28, and 61.)

G. R. Lindsey. 1963. "An Investigation of Strategies in Baseball." *Operations Research* 11:477–501.

J. Thom and P. Palmer, with D. Reuther. 1985. *The Hidden Game of Baseball.* Garden City, N.Y.: Doubleday.

The Consumer Price Index*

Philip J. McCarthy *New York State School of Industrial and Labor Relations, Cornell University*

Almost everyone worries about the prices of all kinds of things. Perhaps the best way for an individual to find out how the price level changes is to follow in newspapers the Consumer Price Index (CPI) of the Bureau of Labor Statistics, a part of the U.S. Department of Labor. Consumers see the CPI as a good measure of price changes of the goods and services they purchase. They react to newspaper statements such as "In September of 1986, the average urban family spent $33.02 for the same amount of goods and services that could be obtained for $10.00 in 1967." They wonder whether income has increased sufficiently to compensate for this increase in prices.

Unions and management pay particular attention to the CPI because they know its value will play a critical role in wage agreements and that a 4% annual

*The original article has been updated by staff at the U.S. Bureau of Labor Statistics.

increase, for example, in the CPI will lead to a demand for at least a 4% increase in wages. Furthermore, in January 1986 there were 4.5 million workers whose wages were covered by contracts containing escalator provisions, that is, provisions calling for automatic changes in wage rates in accordance with specified changes in the CPI.

Americans who get Social Security benefits watch changes in the CPI because it is used to escalate their pensions. The CPI is also used to index payments in other government programs, such as food stamps. The CPI affects government revenue too because it is used to adjust taxable income brackets in our income tax law.

Finally, economists use the CPI as one of the principal indicators to judge the performance of the economy. As Figure 1 shows, the extent of inflation has varied considerably over time.

Although statistical studies of prices and living conditions in the United States were conducted in the late nineteenth and early twentieth centuries, the first complete "cost-of-living" indexes were published by the Bureau of Labor Statistics in 1919. They covered 32 large shipbuilding and industrial centers and arose through an agreement between the Shipbuilding Labor Adjustment Board and labor leaders that one of the factors to be considered in settling labor disputes was that of "adjusting wages to the higher cost of living resulting from the war."

Since that time, the CPI has been broadened in scope and increased in importance. This had led to many professional appraisals of the CPI, among them one by an advisory committee of the American Statistical Association in 1933–1934 and one by the National Bureau of Economic Research in 1959–1960. These appraisals have influenced major revisions in the CPI, made at approximately 10-year intervals since 1940, as well as many minor revisions.

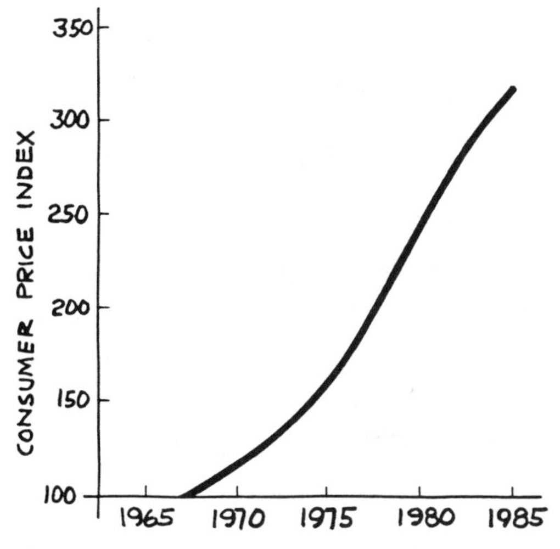

Figure 1 *Consumer Price Index for all urban consumers (CPI–U), annual averages (1967 = 100).*

PROBLEMS IN CONSTRUCTING THE CPI

Typical consumers are quite aware of changes that occur in the prices of goods and services purchased regularly. They know when the price of gasoline increases by one cent per gallon, or when the price of milk is increased by one cent per quart. Furthermore, they are able to predict with reasonable accuracy the impact of these price changes on monthly or yearly budgets and are therefore in a position to estimate the extra income (including an allowance for taxes) needed in order to compensate for these increases. Price changes relating to less frequent and more sporadic purchases—clothing, doctors' services, appliances, automobiles, and homes, for example—are not as visible to the individual consumer, and it is much more difficult to assess their effects on the budget. Thus one may well know that it costs more to live, or less to live, during the current year than in the preceding year, but would find it almost impossible to provide an exact value for this change in the cost of living. This problem would be further complicated by changes that might occur in such matters as family status (births and deaths, divorces), family desires (large cars versus small cars), and family purchasing patterns (chicken instead of steak). And yet it is exactly this evaluation—for the nation as a whole, for separate regions and cities of the nation, for different classes of expenditures (food, clothing, housing, transportation, and the like), and on a monthly basis—that the CPI provides. This essay deals primarily with the contribution of statistics to the construction of the CPI.

If consumers actually wished to keep sufficiently detailed records to measure the change in prices of the items purchased from, say, January 1986 to January 1987, they might proceed as follows. During January 1986, they would keep a record of every purchase and would then summarize these purchases in terms of quantities (number of quarts of milk, number of pairs of shoes, number of haircuts, and the like) and the unit price of each item. In effect, they would define a "market basket" of goods and services and obtain the cost of this market basket in January 1986. Suppose this cost is $750. In January 1987, exactly the same market basket of goods and services would be priced. Let us assume that the cost then was $800. Therefore this consumer's CPI for January 1987 with January 1986 weights (quantities) and with January 1986 as a base (= 100) is $100 \times 800/750 = 107$. The CPI is effectively computed in the same way, but the operation is much more complicated because it deals with a large population rather than a single consumer.

Even if we assume that the family status and desires of our consumer did not change from January 1986 to January 1987—an assumption that can seldom be exactly true—many practical problems and tantalizing questions would plague those responsible for the construction and interpretation of this index. In particular:

1. Even for a single consuming family, the number of different items purchased during a month or year is extremely large, and the continual pricing of such a list would be costly and time consuming. Preparation of a CPI for groups of families obviously magnifies this problem to a wholly impractical size.

2. The average family has available to it a host of vendors of goods and services, and prices vary from vendor to vendor. In January 1987 should one attempt to return to the same places where purchases were made in 1986? And what about stores that have gone out of business or stores that have come into existence in the interim?
3. Prices vary from day to day. Must our consuming family return to the same store at the same time to compute the cost of its market basket in January 1987? And how does one take into account the facts that some families take advantage of sales and others do not and that some families shop on a day-to-day basis while others shop less frequently?
4. Suppose that a major purchase, such as an automobile, was made by our family in January 1986. The fixed market basket still contained this item in January 1987, even though such a purchase was not made at that time. How should this be handled?
5. The goods and services available to the consumer change from time to time. Items contained in the original market basket may not exist in the stores when we return to price them at a later date, or they may exist only at an improved or lowered quality level. What do price changes mean under these circumstances?
6. We also observe that our consuming family—even though its composition, status, and desires are assumed not to have changed—may be able to "beat" a rise in prices by appropriately altering its purchasing patterns. Suppose, for example, that they notice that the price of steak has risen sharply and decide that the desires of the family are as well satisfied by chicken as by steak. If chicken is substituted for steak, the family's satisfaction level will not change, and yet the amount spent for food may actually decrease in spite of what may be a general increase in the price level. In effect, we can think of replacing the "fixed-contents" market basket with a "fixed-satisfaction" market basket.

The above problems relate to both measurement and concept. Thus we may find that a particular item of merchandise has changed in quality, and yet it may be a most difficult task to measure this change in quality and to translate it into a dollar value. The Bureau of Labor Statistics does attempt to account for these changes in quality in constructing the CPI.

Item 6 above is at a deeper conceptual level. Most economists would prefer a "constant-utility" or "constant-satisfaction" or "welfare" type of price index to the "fixed-market-basket" type of index currently produced by the Bureau of Labor Statistics. In other words, what change in expenditures must a consuming family make from one time to another in order to maintain a constant level of satisfaction, with the recognition that the contents of the market basket can be changed in order to accommodate changes in prices, changes in products, and the like? A number of researchers have constructed "constant-utility" indexes, but the practical and conceptual problems still prevent production of such a measure as an official monthly statistical series. As a matter of fact, in September 1945, the official title of the index was changed from the Cost of Living Index to the Consumer Price Index in order to emphasize the distinction between these two approaches. In this essay I shall not treat these conceptual problems but rather focus on the role of statistics as it helps to solve CPI measurement problems.

SAMPLING AND THE CPI

Many of the above problems become more manageable if attention is shifted from the individual consuming family to a group, or *population,* of consuming families. Thus even though some members of the population did purchase automobiles in January 1986, other members of the population did not make such a purchase, and similarly for January 1987. An automobile can then be introduced into the market basket with a weight that will reflect the effects of changes in its average price, averaged over the population of families. This shift in emphasis from the individual consumer to a population of consumers, where differences among the members of this population are known to exist, means that the construction of the CPI requires statistical *sampling* (and analysis) from the population of consuming families.

Moving in this direction also forces one to think in terms of the *population of goods and services* available to all members of the population of consumers and in terms of the *population of outlets* at which all of these goods and services may be purchased. Furthermore, it is manifestly impossible to study every consuming family and to price each item of consumption in all the outlets where it can be purchased. Hence it becomes necessary to select samples from each of these populations and to draw inferences from the samples to the entire populations. The methods of statistics can assist and guide these steps.

The Consumer Price Index has never attempted to measure the changes in the prices of goods and services for all families and individuals living in the United States. Rather, because of the CPI's traditional use in collective bargaining between labor and management, its scope has been restricted to urban families. Until 1978, the population of consuming families covered by the index consisted of all urban families (including single workers) for which 50% or more of the family's income came from wages or from salaries earned in clerical occupations and for which at least one member of the family unit worked for at least 37 weeks during a year. Since 1978, the Bureau of Labor Statistics has produced a CPI for *all* urban consumers (CPI–U) as well as the continuation of the more limited CPI for urban wage earners and clerical workers (CPI–W).

In brief, a sample of urban communities is selected first, and then a sample of families is taken from each of the selected communities. In 1986, this was done in 91 areas. The 29 largest cities in the continental United States, plus 1 city from Alaska and 1 from Hawaii, were automatically included in this sample of cities. The remaining urban areas with populations of 2,500 or more were placed in homogeneous groups according to size and geographic location, and 60 cities were chosen from these groups in accordance with probability methods of selection. This sample of 91 cities serves not only as a basis for studying the expenditure patterns of consuming families but also as a basis for the prices that must be used to determine the current cost of the CPI's market basket of goods and services. The CPI will continue to be based upon this sample of 91 cities until the next major revision takes place.

Within each of the 91 chosen cities, a sample of consuming units was selected and interviewed. These units were drawn in accordance with the tenets of statistical sampling theory; the goal was to choose samples in such a way that objective measures could be devised for assessing the likely size of deviations between averages from the sample and corresponding (unknown) averages from the whole population. This particular survey was called the Consumer Expenditure Survey (CES) because its primary purpose was to collect data relating to family expenditures for goods and services used in day-to-day living.

The CES is really two separate surveys. One is an interview survey conducted at quarterly intervals with each consumer unit in the sample. This survey collects comprehensive data on expenditures, assets, liabilities, and incomes and is most effective for measuring expenditures for large, infrequent purchases such as appliances and automobiles as well as for other household expenses. The second survey is a diary in which each consumer unit in another sample keeps a complete record of all expenses for a two-week period. This method is most effective for measuring small purchases such as hamburger, fish, toothpaste, and photographic film. Each year, 5,000 consumer units are surveyed in each of the two CES surveys. In January 1987, data from the CES for the years 1982–1984 were incorporated into the CPI.

DATA COLLECTION AND THE CPI

Data collected in the CES interviews are used for a variety of purposes in determining the structure of the current CPI. The data define the complete contents and total cost of each family's market basket in the survey years. The items are classified into five major groups: food and beverages; housing; apparel and upkeep; transportation; and medical care, entertainment, and other items. Each of these groups is further subdivided to a final classification scheme consisting of 69 expenditure classes. Some examples of these expenditure classes are: beef and veal, eggs, piped gas and electricity, housekeeping services, footwear, auto repairs and maintenance, hospital and related services, and reading materials. The expenditure classes are further divided into 207 item strata. That is, the consumer's market basket is divided into 207 compartments. The total cost of each compartment is then determined for all sample consumers selected from within a particular city, and these total costs are expressed in relative terms; for example, at December 1986 prices, beef and veal are estimated to account for 1.013% of the total cost of the nationwide market basket. Different market baskets are used in different cities to allow for differences in such characteristics as climate and the availability of different foods. Thus we use many average market baskets rather than a market basket that applies to a single consuming unit.

As noted earlier, the CES interviews define a market basket filled with all the items that consumers purchase. It would be impossible to price all of these items in their almost infinite variety each month, even recognizing that this pricing need be carried out only in the 91 index cities. Hence another sampling problem arises. Again the theory of statistical sampling was used to select a sample of items from each of the 207 item classes. The sampling approach allows the contents of the market basket to change. For example, if one item disappears from the market, replacements may be made by further sampling. The 207 compartments and their original weights, however, remain fixed through time until a new market basket is put in place.

Selecting the specific 120,000 items and identifying the outlets from which prices will be collected is a complex process of several steps. First, many of the 207 item strata or compartments (the fixed building blocks of the index) are further divided into "entry-level items" or ELIs. For example, the item stratum for home laundry equipment is divided into two ELIs: washers and dryers. For each of the urban areas where prices are collected, a sample of ELIs is selected within each item stratum. At least one ELI from each item stratum is priced in each urban area. And over all urban areas, each ELI is priced somewhere, the frequency of its pricing being proportional to its importance within the item stratum. For example, washing machines may be priced in Poughkeepsie and dryers in Grand Island, but nationwide, the proportions of washers and dryers in the CPI are approximately equal to the relative proportions bought by all urban consumers.

With the sample of ELIs in hand for each urban area, we pick a scientific sample of stores or other outlets for pricing each of the ELIs. To do this, the Bureau of Labor Statistics uses a Point of Purchase Survey (POPS) in each urban area. This survey is like the CES in that it determines how much consumer units spend for different classes of items, but it also finds out how much they spend at each of the places from which the items were bought. From this survey, the Bureau constructs what statisticians call a *sampling frame.* For each item category, the sampling frame contains a list of the outlets from which the item category was purchased and of the amounts spent by all sampled consumer units in that outlet. A probability sample of outlets is then selected for each item category, with the result that the sample is representative of all the various types and locations of outlets from which consumers in that urban area purchase items in a category.

Now the Bureau's field staff can visit each of the selected outlets; but which washing machine will be priced? There may be *dozens* of different brands and models. For the CPI the sample must be representative of all different kinds of washing machines. To meet that objective, Bureau staff and store personnel review the dollar volume of sales for each type of machine and then select a probability sample from the ones sold in that store. The greater the sales of a particular variety, the greater the probability that it will be included in the sample, but every variety—no matter how obscure—has some chance for selection. The final result is that the national sample for each item stratum in the CPI is composed of many different varieties of an item. The varieties on which consumers, on average, spend more, appear more frequently in the sample list while those for which less is spent appear proportionately less frequently.

The field operations of monthly pricing are intricate. Not only must a large field staff be trained and supervised, but the staff must get the cooperation of store managers and provide for businesses that cease operations or come into existence during the 10-year period that ordinarily elapses between major revisions. To appreciate the magnitude of these endeavors, note that prices are collected at 21,000 outlets each month.

The CPI has a final sampling problem, not always recognized as such. Although the CPI is published monthly, it does not refer to any definite date during the month. The pricing operation has to be almost continual, and it is therefore necessary to choose a sample of times when prices are obtained. This sampling is not carried out in as formal a manner as are the other sampling operations. Nevertheless, every attempt is made to ensure, for example, that sale and nonsale days for food are represented in their proper proportions and that a similar balance is maintained for other items, such as newspapers and theater admissions, whose prices may change periodically.

One important goal in statistical design and analysis is to have an objective measure of the precision of sample analyses. Although this goal has not yet been fully realized in the complex setting of the CPI, substantial progress has been made. All indications are that the sampling operations are reasonably well under control and that uncertainty in the value of the index due to sampling is relatively small compared to the uncertainties arising from other aspects of the process, for example, the effects of quality changes on the index.

SUMMARY

The production of monthly values for the Consumer Price Index by the Bureau of Labor Statistics is a highly complex undertaking that involves problems of *basic economic theory* (for example, the choice between a price index or a constant-satisfaction index), *measurement and quantification* (for example, of changes in the quality of items purchased by consumers), *sampling statistics* (definition and selection of samples from a wide variety of populations), and *operations* (for example, training and supervision of price reporters).

The CPI is concerned with a population of consuming families and with the population of cities in which these families live. A sample of cities serves two purposes. First, from within the selected cities a sample of consumers is chosen from which it is possible to determine average expenditure patterns. Second, prices are collected in the selected cities to determine the value of the current CPI. The Consumer Expenditure Surveys also define a population of goods and services for which consuming families spend their income. From this population a sample must be taken for current pricing purposes. Within each of the index cities there exist populations of outlets at which items can be purchased, and samples must be chosen to represent these populations.

Finally, there is a population of times within a month at which price quotations can be obtained, and this population also must be sampled.

It is certainly true that everyone may not be completely satisfied with the CPI in its present form. Improvements can and probably will be made in many parts of the index, for example, in the basic data on consumer expenditure patterns, in the sampling of outlets, and in the collection of price data from these outlets, in the preparation of indexes for a wider variety of subpopulations, and in techniques for handling quality change problems. It may even happen that some completely new approach to the construction of the index may be developed, possibly through the use of newer mathematical techniques. No matter what its form, however, the CPI undoubtedly will remain one of the main indicators of the state of the U.S. economy.

PROBLEMS

1. Refer to Figure 1. In 1980 how much would the average urban family have paid for the same amount of goods and services that it could have obtained for $10 in 1967?
2. Explain the difference between a "constant-utility" and a "fixed-market-basket" type of index. Which type would you prefer the Bureau of Labor Statistics to use? Why?
3. What is the purpose of dividing up cities into homogeneous groups before choosing the cities to be included in the price survey?
4. Why are all of the largest cities automatically included in the price survey?
5. Suppose a group of rural members of Congress push through a bill requiring the Department of Agriculture to find out the cost of living for farm families.
    a. Could the department refer them to the CPI? Why or why not?
    b. If not, what steps would the department have to take to implement this law?
6. Suppose an anthropologist studying a remote community of Northern Albertan gold prospectors decides to construct a CPI for the community. A survey in January 1975 finds that the prospectors spent on the average:
    a. 100 ounces of gold for flour, which costs 4 ounces a sack;
    b. 20 ounces of gold for burros, which cost 100 ounces on the average;
    c. 40 ounces of gold for whiskey, which costs 5 ounces per pint.

In January 1976, the anthropologist finds that flour has gone up to 6 ounces of gold per sack; burros sell for only 80 ounces on the average; and whiskey costs 10 ounces a pint. What is the CPI for January 1976 (January 1975 = 100)?

REFERENCES

Ethel D. Hoover. 1968. "Index Numbers: II. Practical Applications." In *International Encyclopedia of the Social Sciences,* vol. 7, David Sills, ed. New York: Macmillan and Free Press, pp. 159–165.*

Philip J. McCarthy. 1968. "Index Numbers: III. Sampling." In *International Encyclopedia of the Social Sciences,* vol. 7, David Sills, ed. New York: Macmillan and Free Press, pp. 165–169.*

Erik Ruist. 1968. "Index Numbers: I. Theoretical Aspects." In *International Encyclopedia of the Social Sciences,* vol. 7, David Sills, ed. New York: Macmillan and Free Press, pp. 154–159.*

U.S. Department of Labor, Bureau of Labor Statistics, 1984. *BLS Handbook of Methods: Vol. II, The Consumer Price Index.* Bulletin 2134-2. Washington, D.C.: U.S. Government Printing Office.

*The first three references are also available in William H. Kruskal and Judith M. Tanur, eds. 1978. *International Encyclopedia of Statistics,* vol. 1. New York: Macmillan and Free Press, pp. 451–467.

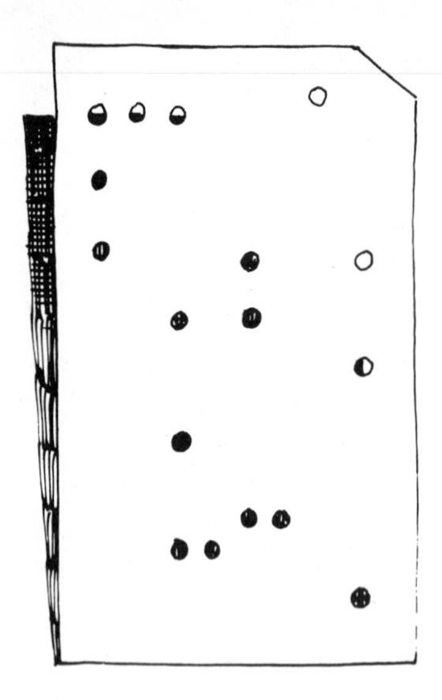

How To Count Better:

Using Statistics to Improve the Census

Morris H. Hansen *Westat, Inc.*
Barbara A. Bailar *Bureau of the Census**

THE IMPORTANCE OF CENSUS RESULTS

In 1790, Thomas Jefferson gave to George Washington, our first President, the results of the first census of the United States of America. Every 10 years since then, as provided in the Constitution, the decennial census has determined for the nation essential information about its people.

The basic constitutional purpose of the census is, of course, to apportion the membership of the House of Representatives among the states. From the beginning, the census has had many important purposes beyond the constitutional one. The development of legislative programs to improve health and education, alleviate poverty, augment transportation, and so on, is guided by census results. They are used also for program planning, execution, and evaluation.

*Now with the American Statistical Association.

Now the distribution of billions of dollars a year from the Federal Government to the states, and from the states to local units, is based squarely on census counts. Private business uses the census for such purposes as plant location and marketing. Much social and economic research would be essentially impossible without census information.

The importance of the census to current problems such as poverty, health, education, civil rights, and others brought about requests for shifting to a 5-year, rather than a 10-year, census in order to keep information more nearly up to date. Those requests came from governors of states, mayors of large cities, scientists, business people, and many others. In 1976, legislation was passed that called for a 5-year census, but Congress has never appropriated the funds to carry it out.

THE JOB OF PLANNING AND TAKING THE CENSUS

The need for a fast and reasonably accurate census is fairly obvious. What may not be so apparent are the major problems involved in taking a census and making the results available promptly, and in adapting the census questions to serve current needs. In each census, some questions have been changed in response to new needs, but certain basic information has been consistently required. Questions in the 1980 population census included age, sex, race, marital status, *household* relationship, education, school enrollment, employment and unemployment, occupation and industry, migration, *travel time to work, persons in carpool,* country of origin, *current language and English proficiency, ancestry,* income, and other subjects. (Items in italics were new in the 1980 census.)

The job of organizing and taking the census is a major administrative and technological undertaking. Even though most of the questionnaires are filled out by the respondents themselves in a "census by mail" and modern computers and other advanced technology have eliminated a lot of paperwork, the taking of a census requires the recruiting and training of hundreds of thousands of people, most of them for only a few weeks of work. The 1980 census, for example, recruited some 458,000 persons, with 270,000 persons active during the peak period of operations. Once specific goals are set in terms of questionnaire content and desired statistical results, the massive job of organization and administration begins.

The system for canvassing and for collecting, receiving, processing, and summarizing the vast numbers of completed questionnaires must be planned. Specialized electronic and paper-handling machines designed and built at the Census Bureau automatically read the information recorded by respondents or enumerators. These complex machines first photograph on microfilm and then scan and read microfilm copies of the census forms that have been filled out, in most cases, by the respondents themselves. The magnitude of the job is difficult to comprehend. For the 1980 census, a little more than a quarter of a billion pages (counting each side of a relatively large sheet as a page) were handled. After being clerically reviewed, the results were recorded on magnetic tape, and then computers examined these results. In the process, the forms were edited for certain types of incompleteness or inconsistency, and adjustments were made automatically or special problem areas were identified for further manual investigation. The later steps of tabulation and printing for publication also were accomplished on electronic computers.

The approach to the job of taking a census differs from that of designing a totally new system, in that the census has been taken many times before, and the background and experience of the past serve to guide current efforts.

The availability of extensive past experience, however, has disadvantages. There may be long traditions that have come to be regarded as essential but that, in fact, only represent ways in which the job has been done in the past. For example, the tradition of taking the census by an enumerator canvassing an area and personally asking the questions of any responsible member of the household had been long established. This approach was regarded as proven by long use to be the only reasonable way to take the census. Because the concepts in some census questions are difficult, it was thought that only a trained enumerator could ask the questions and elicit the proper information. But this view did not recognize the difficulties in training and controlling an army of temporary interviewers. Nor did it recognize that the responses obtained in the census interview situations were sometimes based on a misunderstanding of the questions and that the interview process did not allow time for a considered reply. Furthermore, the interviewers' conceptions or misconceptions could importantly influence the response.

In the nineteenth century, many potential advantages of prior experience were lost, for the Census Bureau was not created as a permanent and continuing agency until 1902. It then became far more feasible, with a continuing staff, to benefit from lessons and experience of prior censuses. The situation for the 1900 and earlier censuses is illustrated in the following quotation from *The History of Statistics* (Cummings, 1918, pp. 678–679).

> Mr. Porter gives the following account of his experience, which must have been essentially that of every Superintendent of the Census.
>
> The Superintendent in both the last two censuses [1880 and 1890] was appointed in April of the year preceding the enumeration, but when I was appointed I had nothing but one clerk and a messenger, and a desk with some white paper on it. I sent over to the Patent Office building to find out all I could get of the remnants of ten years ago, and we got some old books and schedules and such things as we could dig out. . . . I was not able to get more than three of the old men from the city. . . . I knew most of the old census people. Some of them were dead and some in private business. . . . But little over sixty days were allowed for the printing of 20,000,000 schedules and their distribution, accompanied by printed instructions to the 50,000 enumerators all over the country, many of them remote from railroads or telegraph lines. . . . Now to guide us in getting up these blanks, we had only a few scrapbooks that someone had had the forethought to use in saving some of the forms of blanks in the last

census. He had taken them home, a few copies at a time, and put them into scrapbooks. The government had taken no care of these things in 1885, when the office was closed up. Some of them had been sold for waste paper, others had been burned, and others lost.

In addition to showing the potential gains from continuity and learning from the past, this quotation suggests the great complexity of the census in the latter half of the nineteenth century. At that time there was little in the way of a continuing statistical program in the Federal Government and as a consequence the decennial census was loaded with a range of questions that proved difficult if not impossible to collect through decennial census inquiries—hence the great number and variety of questions and forms. Many of these types of information are now collected through sample surveys or compilations from administrative records.

THE USE OF STATISTICS IN PLANNING AND TAKING THE CENSUS

Statistical concepts and methods have provided fundamental improvements to the census over the past 50 years. We might say that these improvements form a technological explosion. Part of this explosive change has come from the introduction of the large-scale electronic computer, but even more of the change has come from statistical advances and the application of statistical studies. (In fact, development of the high-speed computer and modern statistical methods were both substantially motivated by census problems and were in part carried out by census scientists or with Census Bureau support.)

The Introduction of Sampling as a Tool for Census Taking

Sampling was first used in collecting census information in the 1940 census. A series of questions was added for a 5% sample of the population. [Roughly speaking, this meant asking every twentieth person the additional questions (Waksberg and Pearl, 1965).] This was a major advance, as the tradition for a century and a half had been universal coverage for every question. The questions asked of the 1940 sample included one on wage and salary income—income had not previously been a census question—one on usual occupation (as distinguished from current occupation; the 1940 census was planned during the Depression, when unemployment was very high and there was frequently a difference between a person's usual occupation and the occupation at which he or she was currently working), and several other questions.

The art and science of modern statistical sampling were evolving at the time of the 1940 census, and at the same time there was increasing public acceptance of sampling. It was possible to proceed with greater knowledge and confidence about what a sample would produce than would earlier have been possible. Thus, for a particular size and design of sample, statisticians could establish a reasonable range for the difference between a sample result and that obtained from a complete census. Suppose, for example, that a city had a total population of 100,000 persons of whom 30,000 were employed and earned wages or salaries, and suppose that 10,000 of these received wages and salaries of less than $2,000 in 1939 (the year preceding the census). There would be approximately 5,000 people in a 5% sample from that city. The estimate of the number receiving less than the $2,000 wage and salary income would, with a very high probability, when estimated from the 5% sample, lie within the range 9,300 to 10,700. This kind of accuracy was sufficient to serve many important purposes and, in fact, was as much accuracy as could be justified in the light of the less-than-perfect accuracy of the responses to the question on income. Not only could estimates be prepared from the sample of what would have been shown by a complete census, but, in addition, the range of probable difference between the sample estimate and the result of a complete census could be estimated from the sample! Sampling theory also guides in designing samples to achieve a maximum precision of results per unit of cost.

In considering the advantages of the use of sampling, it may appear to some that the main work involved in taking a census is the time it takes in going from door to door and that, once some questions have been asked at a household, the cost of adding questions would be small whether they were added for a 5% sample or for the total population. Such a presumption is far from true. Suppose, for example, that an additional question about whether a person has a chronic illness adds an average of 20 seconds of work for each person counted in the census. With some 200 million people in the population, this would add more than 1 million hours of work and perhaps $4 million to the cost of the census. Thus obtaining the added information for, say, five questions from only a 5% sample instead of from all persons can produce needed and highly useful results at a fraction of the cost for complete coverage. Finally, the use of a sample permits tabulation and analysis much sooner than complete coverage does.

Starting in the 1940s sample surveys on a wide range of subjects were introduced so that continuing and up-to-date information would be available between the censuses. For example, the Current Population Survey is a sample of the population conducted monthly by the Bureau of the Census. The Survey collects information each month on employment, unemployment, and other labor force characteristics and activities of the population. It also serves to collect information on other subjects, with different supplemental questions in various months. In one month each year almost the full range of population census questions is asked. In other months information may be requested on recreational activity, housing, disability, or other subjects. (See the essay by Leon and Rones for further discussion of the Current Population Survey.) Sample surveys cover health, retail trade, business and personal services, the activities of governmental units, and so on. These surveys have large enough samples to provide national information, some information for regions of the nation, and even information for large states and metropolitan areas. They cannot, however, provide information for the many individual cities and counties and for relatively small communities within the cities and counties. To obtain that kind of fine detail, very large samples are needed, samples such as those taken as part of the decennial census.

In the 1950 census, the use of sampling was extended to some questions that in earlier censuses had been collected from all persons. For these questions, a 20% sample was used. This sampling was extended in the 1960 census to most of the items of information. In the 1960, 1970, and 1980 population censuses, only the basic listing of the population, with questions on age, sex, race, marital status, and family relationship, was done on a 100% basis.

The following question is often raised: If sampling is so effective a tool, why not take the whole census with a sample? Isn't it a waste of effort to do a complete census? One must remember that a primary purpose of the census is not to obtain national information but to provide information for individual states, cities, and counties, and for small areas within these. The results obtained by converting the whole census to a relatively large sample (perhaps including 20% of the population) would be adequate for some purposes. Such a sample, however, would not apportion representatives in the state legislatures in the same way as a complete census. Similarly, there could be important differences in the distribution of vast amounts of funds to thousands of individual small areas. Also, in some states, the legal status of many communities depends on the exact size of the official population count; for example, a city with 10,000 or more people can issue bonds. Hence a complete census is needed for total population counts and for some other basic population data, but the great bulk of the information may be collected from a relatively large sample.

Income—An Example of Sampling in the Census

When the 1940 census was planned, after a long depression, questions on wage and salary income were put on the census questionnaire to guide the nation on government and private sector programs. The income question was asked of everyone in the census, and a storm of protest was raised. There was some congressional opposition, and a New England senator led a campaign to persuade the public not to respond. Issues of privacy and confidentiality were given much play by the press. Congressional opposition and newspaper editorials focused on these issues of invasion of privacy and increased government intervention. There was support from the administration to ask the question. To allay fears of lack of privacy, the Census Bureau printed 20 million forms that individuals could fill out and mail directly to Washington, so that no enumerator could see the response. Only 200,000 such forms were ever used, and the nonresponse rate was less than 2%.

Since 1940, however, income has been asked routinely on census questionnaires for a sample of the population. In 1950 the expanded income questions were asked only of a 25% sample of the population. The practice of sampling for income questions has continued but it caused a problem after the 1970 census. Revenue sharing began in 1972, allocating billions of dollars of Federal funds to state and local governments, based on formulas that included local income estimates. For small areas, the 20% sample used in 1970 gave too much variability in sampling error between smaller and larger places, sampled at the same rate. Thus, in 1980, two sampling rates were used. A one-in-two rate was used for counties, incorporated places, and minor civil divisions with a population of 2,500 or less. The rest of the country was sampled at a one-in-six rate. This differential sampling satisfied the data needs for small areas.

The uses of income data are so sweeping that it is now seen as one of the most important census items. There are over 20 Federal programs where the use of census data on income is legislated.

The Use of Sampling and Experimental Studies to Evaluate and Improve Census Methods and Results

Substantial steps to evaluate and improve census methods begun in the 1940s have been continued and greatly extended since then. Statistical studies have been made of various aspects of the census. One such study was made by repeating the census enumeration, in a well-designed sample of areas, shortly after the original census enumeration, using the same procedures as in the initial census. Thus we were able to see something of how much alike two censuses taken under the same conditions and procedures would be.

Studies of these types show high consistency and accuracy of response for questions such as sex, age, race, and place of birth, but they show higher degrees of inconsistency and inaccuracy in responses to the more difficult questions relating to occupation, unemployment, income, education, and others. The information from such studies helps both in improving the questions in the next census (by, for example, showing which questions cause trouble and need rewording) and in interpreting the accuracy of census results when the questions are put to specific use.

Studies that compare alternative methods and procedures within the framework of well-designed and randomized experiments have been exceedingly important for learning about the effectiveness of various procedures and in comparing their cost and their accuracies. Such comparisons have been made, for example, of various types of questionnaire design and other variations in procedures. For example, one study reexamined the census coverage and questions in a sample of areas, but used more highly trained enumerators, more detailed sets of questions, and other such expensive improvements.

These rather wide-ranging studies led generally to the conclusion that some of the methods earlier regarded as the way to achieve major improvements would not be effective in relation to their cost (although some worthwhile improvements in questionnaire design and procedures were accomplished). We find, for example, that simply spending much more time and money on training an army of temporary enumerators would add to cost but probably not lead to corresponding improvements in accuracy. A problem in the 1970 and 1980 censuses has been the turnover of enumerators. Many are trained but some quit before starting work and others do the easy cases and then quit. Supplemental pay experiments in the test censuses leading to 1990 have been encouraging. Employees who meet standards qualify for supplemental payments. In the test censuses, the increase in production and the resulting savings in enumerator wages have more than compensated for the cost of the supplemental payments. Continued experimentation is planned before the final decision on 1990 pay plans.

The Surprising Effect of Enumerators

One study, however, led to surprising conclusions and then to a basic improvement in census procedures. It had long been known that enumerators can and do influence the answers they obtain—presumably unconsciously most of the time. But the magnitude of this enumerator effect was not known. Hence a large statistical study was carried out as part of the 1950 census to measure the magnitude of enumerator effect.

The study plan (in a simplified description) was based on areas divided into 16 work assignments (areas small enough so that 2 work assignments could be canvassed by a single enumerator) and 8 enumerators. Two of the 16 work assignments, chosen at random, were given to the first enumerator; two of the remaining ones, also chosen at random, were given to the second enumerator, and so on. Of course, the whole 16-fold experiment was repeated many times.

The random choice of work assignments was essential here in order to interpret the results in a useful way. (Random choice means choice by a method equivalent to writing the numbers 1, 2, . . . , 16 on identical cards, shuffling or mixing them thoroughly, and then picking first one, then another, and so on.)

Another essential feature of the plan was that each enumerator had 2 work assignments and that there were a number of enumerators. That way, good estimates could be obtained for the variability introduced by a single enumerator (roughly, the differences in performance by the same enumerator that were not attributable to differences in areas) as compared with the variability stemming from differences among the enumerators.

Further details of this path-breaking analysis cannot be given here, but we can summarize the results. Far greater differences between enumerators were found than had been anticipated, not so much on items such as age and place of birth, but on the more difficult items such as occupation, employment status, income, and education. For those items, in fact, a complete census would have as much variability in its results (because of enumerator effect) as would a 25% sample if there were no enumerator effect!

What should we do about this? One approach might be to expend far more resources on the selection, training, and supervision of enumerators. But the other studies mentioned above have shown that this is not feasible under the conditions of a national census.

Another possible answer is to eliminate the need for the enumerator by leaving the carefully designed questionnaires with the respondents and asking them to fill them out and mail them. (Enumerators would be needed only when respondents' returns were incomplete or where the respondents asked for help.) This method was tested in further studies and found to work quite well. It was used in the 1960 census for the longer census forms and was a great success, in terms of both cost and added accuracy. Hence, in 1970 and 1980, this method was used still further: most of the population received and returned the census forms by mail.

Thus, in the process of using statistics to improve the census, the completeness of coverage has been improved, and the accuracy of the items of information collected has been increased.

COVERAGE ERROR IN THE CENSUS

Since 1950, the Census Bureau has been measuring the coverage it attains in the census. The rates are very high, between 98% and 99%. However, the rates vary over population subgroups. About 5% of the Black population is missed, with rates approaching 20% for Black males in certain age groups. Hispanics are also counted at lower rates.

After the 1980 census, the Census Bureau was sued by several localities. Because of the concentration of the undercount in minority groups, cities with large numbers of minorities were convinced that they had been undercounted. They asked for a statistical adjustment of the census results. In the court cases, the Census Bureau argued that though statistical methods existed that were accurate enough to measure the undercount for evaluation purposes, the methods were not accurate enough to adjust the census counts. Other statisticians, representing the jurisdictions affected, said the population distributions would be improved by adjustments. One of the lawsuits—that filed by the state and city of New York in 1980—was decided after seven years. On December 8, 1987, Federal Judge John Sprizzo issued an opinion in favor of the government. At the time this essay was written, however, many other suits remain to be decided.

The Census Bureau instituted an intense, focused research program on methods of measuring the undercount and distributing it to all levels of geography. Several demonstrations have taken place in test censuses. However, on October 30, 1987, the Department of Commerce announced that it had decided against making an adjustment for any undercount for 1990.

PROBLEMS

1. Give several reasons why a census is taken.
2. Why is it desirable in some instances to take a 5% sample as opposed to a complete census? What does the 5% sample lose?
3. Comment on the advantages and disadvantages of using enumerators in a census.
4. Answers to questions regarding sex, age, race, and place of birth seem to be more reliable than those regarding occupation, unemployment, income, and education. Give two possible reasons for this. Do you think answers to mailed questionnaires would be more accurate than interviews by enumerators on these questions? Why or why not?
5. (For group work.) Take a small random sample of people and administer a questionnaire on some issue, for example opinions of the local newspaper. Report on the problems and situations that arise.
6. Suppose the Census Bureau decided to add a battery of six questions on energy consumption to the 1990 census. About how much money would

the Census Bureau save if, instead of asking everyone the questions, they asked them of only a 1% sample?

7. Suppose that a census uses mailed questionnaires instead of enumerators, and 40% of the returned questionnaires have no answer marked to a question on mental health. Would the Census Bureau be justified in counting only the marked responses and publishing the result as a 60% sample? Justify your answer.

8. A scientist in the Census Bureau suggests using telephone interviews instead of mailed questionnaires for families with telephones, thinking that this technique might lead to a more accurate census.
   a. What arguments can you think of which support or refute this idea?
   b. Design an experiment to determine whether this idea is correct.
   c. What other considerations (beside accuracy) would determine whether or not the telephone technique would be adopted?

REFERENCES

Barbara A. Bailar. 1983. "Counting or Estimation in a Census—A Difficult Decision." *1983 Proceedings of the Social Statistics Section of the American Statistical Association.* Washington, D.C.: American Statistical Association, pp. 42–49.

Barbara A. Bailar. 1985. "Comment." *Journal of the American Statistical Association* 180: 109–114.

John Cummings. 1918. "Statistical Work of the Federal Government of the United States." *The History of Statistics,* John Koren, ed. New York: Macmillan, pp. 678–679.

Morris H. Hansen. 1987. "Some History and Reminiscences on Survey Sampling." *Statistical Science* 2(2): 180–190.

Morris H. Hansen, Leon Pritzker, and Joseph Steinberg. 1959. "The Evaluation and Research Program of the 1960 Censuses." *1959 Proceedings of the Social Statistics Section, American Statistical Association.* Washington, D.C.: American Statistical Association.

Morris H. Hansen and Benjamin J. Tepping. 1969. "Progress and Problems in Survey Methods and Theory Illustrated by the Work of the United States Bureau of the Census." In *New Developments in Survey Sampling,* Norman L. Johnson and Harry Smith, Jr., eds. New York: Wiley (Interscience), pp. 3E(1)–(11).

Ingram Olkin. 1987. "A Conversation with Morris Hansen." *Statistical Science* 2(2): 162–179.

Ann Herbert Scott. 1968. *Census, U.S.A.: Fact Finding for the American People 1790–1970.* New York: Seabury.

Joseph Waksberg and Robert B. Pearl. 1965. "New Methodological Research on Labor Force Measurements." *1965 Proceedings of the Social Statistics Section, American Statistical Association.* Washington, D.C.: American Statistical Association, pp. 227–237.

How the Nation's Employment and Unemployment Estimates Are Made

Carol Boyd Leon*
Philip L. Rones *Bureau of Labor Statistics, Office of Employment and Unemployment Statistics*

Every month, like clockwork, newspapers around the country publish a headline such as "Employment Continues to Climb," "Unemployment Falls Sharply," or "Unemployment Rate Reaches Record High." Even "U.S. Employment and Unemployment Picture Unchanged" makes the news.

These stories are based on the Bureau of Labor Statistics' monthly report on the nation's employment situation. The report attracts wide attention among policymakers, researchers, business analysts, the news media, and the public at large because the condition of the nation's job market affects everyone.

The news goes out at 8:30 A.M. (Eastern Time) on the first Friday of each month when the BLS issues "The Employment Situation" news release. For the rest of that day, a staff of about 20 Bureau economists is kept busy pro-

*Carol Boyd Leon, formerly an economist with the Bureau of Labor Statistics, is now a free-lance writer in the Washington, D.C., area.

viding data and insights into the newly released numbers. Calls come from the press, financial institutions, foreign embassies, and a host of other sources. Later that morning, the Commissioner of Labor Statistics appears before the Joint Economic Committee of Congress to tell them more about what happened and to answer their questions as TV cameras roll and reporters take notes.

All this happens only after an involved process of gathering data, reviewing survey responses, transferring the information to computer tapes, creating tables, adjusting the data for seasonal fluctuations, and analyzing the figures to discern important movements and trends. This process, from the collection of the data to their appearance on the morning news, takes less than three weeks.

To obtain the data, the Bureau conducts two separate surveys each month. The survey of employment and unemployment dates from 1940, and estimates of payroll jobs go back to 1915 for some industries. The public may be familiar with the results, but how these estimates are made remains a mystery to most Americans.

THE CURRENT POPULATION SURVEY

Your home may not have been visited by an interviewer gathering data for the monthly survey of employment and unemployment, as the Census Bureau, which collects the data for the Bureau of Labor Statistics, interviews fewer than 60,000 households each month. It is hardly surprising that most Americans are unaware of the major effort made—through the Current Population Survey (CPS)—to assure that reliable and timely statistics are collected on a round-the-year basis.

Every month (during the week that includes the 19th), interviewers talk with a scientifically selected sample of households to find out how many people live there and how many were in the labor force—that is, were either working or looking for work—during the previous week, the so-called reference week, the week that includes the 12th. The first question the interviewer asks (after preparing or verifying a roster of household members) is "What were you doing most of last week, working or something else?" If "something else" is the response, the interviewer continues: "Did you do any work at all last week, not counting work around the house?" Any work for pay or profit qualifies one as employed.

If no work is reported, the questioning takes a different course: "Have you been looking for work during the past 4 weeks?" "What have you been doing to find work?" If the respondent has made some effort to find a job, he or she is counted as unemployed.

If there is no job search, the CPS classifies the respondent as not being in the labor force. Interviewers never ask directly, "Are you employed or unemployed?" Instead, the responses to the above questions classify the person on the basis of what they are doing rather than how the person categorizes himself or herself. Thus labor force classification is based more on objective measures of activity than on subjective responses.

Once the interviewer obtains the basic status, she (most interviewers are women) asks additional questions to learn more about each person. For example, employed persons are asked about the hours they worked, the types of jobs they held and the business or industry they worked in. Job seekers are asked how long they have been looking for work and if they started job hunting because they had lost or quit a job, or for some other reason. Persons not in the labor force are queried about their current desire for a job and their intent to look for one in the future.

In addition to counts of persons employed, unemployed, and not in the labor force, the survey also provides BLS with a host of unemployment rates for specific age/race/sex groups, ratios of employment to population, rates of labor force participation, and literally thousands of other bits of information that tell how many people are working, what they are doing, and how many are looking for a job.

HOW RELIABLE ARE THE EMPLOYMENT AND UNEMPLOYMENT ESTIMATES?

Although only a very small proportion of the nation's households are interviewed each month, the sample provides reasonably accurate estimates for the nation's entire population aged 16 and over. The accuracy of the data could be improved by greatly expanding the sample—at the limit by undertaking a monthly census of the entire population like that done every 10 years. The costs of doing so, however—in dollars and in respondent burdens—would be prohibitive. The 1990 Census will probably cost over $1 billion for field operations alone.

Even though the household survey is large enough to ensure adequate reliability, its statistics may differ from those that would be obtained from a complete census. Fortunately we do have some estimates of the accuracy of the CPS.

For example, the overall unemployment rate is generally accurate to within about two-tenths of a percentage point. That means that if the "real" rate for the population were 7.0%, then there is a 90% chance that the results from the survey would fall within the 6.8% to 7.2% range. By comparison, the rate for teenagers, a much smaller group, is accurate to about plus or minus one full percentage point. Because any month's unemployment rate is a statistic subject to error, the difference between two months is also. For a month-to-month change in the unemployment rate to be considered statistically significant, it must be greater than or equal to 0.2 percentage points.

WHAT ARE THE SAMPLING AND ESTIMATION PROCEDURES?

Households are chosen from 729 sample areas spanning 1,973 counties and independent cities. These sample areas—located in every state plus the District of Columbia—include addresses in large and small cities, suburbs, and rural areas.

In translating survey findings into national estimates, each person interviewed gets a weight equal to the number of people he or she represents; each person represents roughly 1,800 people nationwide. Also, because the characteristics of the people selected for the survey may differ from those of the whole population, the survey estimates are inflated to match updated census totals that reflect the characteristics of the population by age, sex, race, and other factors.

The household at each address in the sample is interviewed for 4 consecutive months, omitted for 8 months, and then interviewed again for the same 4 months of the next year before being dropped from the sample. This means that three-fourths of the sample addresses are the same from month to month and half from year to year. Hence, two goals are accomplished: The 4-8-4 pattern ensures considerable consistency in the sample and no household is too burdened by participation in the survey. (The typical CPS interview takes only 15 or 20 minutes.)

After the Census Bureau collects and tabulates the CPS data, it transfers them to the Bureau of Labor Statistics, which is responsible for analysis and publication of labor market data.

What does the CPS tell us? Among many things, we can learn the unemployment rate for black teenagers, the labor force participation rate for married women ages 25 to 34, the employment-population ratio for men ages 65 to 69, the unemployment rate for workers in the automobile industry, and the number of persons on layoff from construction jobs.

We can also learn such things as how many persons not in the labor force want a job but aren't looking for one because they think they can't find one (so-called discouraged workers), the median number of weeks persons of Hispanic origin have been unemployed, the earnings of women who work in service occupations, and how many families with children have no employed member. Indeed, the CPS generates a wealth of information about the labor force. In addition, special questions supplement the regular questionnaire in some months. Every March, for example, the CPS asks about work during the prior calendar year. Other CPS supplements are one-time attempts to obtain data on current topics, such as worker displacement or job training.

Policy needs change over time, creating a need for special statistics, such as data on Vietnam-Era veterans, discouraged workers, or the comparative earnings of men and women. Nevertheless, the basic concepts and definitions used in the CPS have stayed essentially the same. There is also a fierce insistence that the statistics be presented in an objective, factual way that doesn't impart a political view. Everyone at the Bureau of Labor Statistics, from the most junior analyst to the Commissioner, believes that he or she can be of most service when making a straightforward presentation of data that allows policymakers to make an informed decision. Although informed people often disagree about what the numbers mean, most parties appreciate the Bureau's role as an impartial reporter, free from outside political pressures.

It surprises some that the release of rather dry statistics can be the source of so much controversy. We are all sure that unemployment is bad and employment is good. Doesn't publication itself serve to show what is right and wrong with the labor market?

Not really. For instance, economists agree that some level of unemployment is the normal, even healthy, result of people entering or reentering the labor force or voluntarily leaving one job to look for another one. But experts do not agree on how high that healthy level is.

Certainly, unemployment stood at a shockingly high level during the Great Depression, and many suffered severe hardships on account of it. Even in relatively good times, some families lose their homes because a lost job means the mortgage can't be paid. But not every unemployed worker suffers to the same degree. For instance, a student from a comfortable family who casually looks for a job to provide some spending money counts as unemployed just as does a single parent who loses the job that paid the rent. The statistics alone cannot settle policy questions, but they do provide basic facts needed before useful debate can take place.

WHAT ARE THE OFFICIAL DEFINITIONS OF EMPLOYMENT AND UNEMPLOYMENT?

Labor force definitions have been modified, but not substantially altered, since the inception of the CPS. Since 1967, the following basic definitions have been used:

Employed persons are those who did any work for pay or profit during the survey week or who worked at least 15 hours as unpaid workers in a family-owned business. Persons temporarily absent from work because of illness, vacation, or other such reasons are also considered employed.

Unemployed persons are those who had no job at all during the survey week, were available for work, and had made specific efforts to find a job during the prior four weeks. These efforts may have been formal, such as registering with an employment agency or interviewing for a job opening, or rather informal, such as checking with friends or relatives about available work. The Bureau also counts as unemployed persons those waiting to be recalled to a job after being laid off or waiting to start a new job within 30 days.

The CPS unemployment data are not counts of persons receiving unemployment compensation. Although the Department of Labor keeps track of unemployment insurance claims, the so-called insured unemployed comprise less than one-third of the total unemployed. The main reason this share is so low is that many of the unemployed have never held a job or are reentering the labor force after a period of absence. Others are not receiving unemployment compensation because they lost a job that was not covered by unemployment insurance, they had not worked long enough at a covered job, or they had been jobless so long that they exhausted their benefits.

The *labor force* is the sum of the employed and the unemployed. Of the 180 million people age 16 and over in 1986, about 118 million, or 65%, were in the labor force.

The *unemployment rate* is the number of unemployed persons as a percent of the labor force. For example, 7 million unemployed out of a labor force of 100 million yields an unemployment rate of 7.0%. What we hear most about is the unemployment rate, not the actual numbers of employed or unemployed. Indeed, this is one of the most publicized statistics in the country.

In calculating the civilian unemployment rate, only civilians are taken into account, while members of the stateside Armed Forces are included as employed in the rate for the total labor force. The civilian unemployment rate is typically one-tenth of a percentage point higher. The civilian unemployment rate has been as high as 25% and as low as 1.2%.

The Bureau of Labor Statistics recognizes that some believe that the official definition of unemployment is too broad while others think it is too restrictive. To take into account such views, the Bureau publishes rates based on varying definitions of unemployment and the labor force. One includes only those unemployed 15 weeks or longer. At the other extreme, one includes not only the total civilian unemployed but also discouraged workers and persons working part-time for economic reasons.

The *labor force participation rates* tell us what proportion of the working-age population is in the labor force. That rate has grown slowly but steadily since the mid-1960s, primarily because the rapid increase in paid employment among women has more than offset a growing trend among men to retire earlier. On average, about 76% of men and 55% of women were labor force participants in 1986.

The *employment-population ratio* indicates the proportion of the working-age population that is employed. Except during the sharpest business downturns, employment increases. Looking at employment growth in relation to population growth puts changes in the number of workers into a more meaningful perspective.

Economists use these ratios and other analytical tools to examine the current U.S. labor market and labor force trends. The Bureau of Labor Statistics alone has issued some 700 articles and special studies analyzing CPS data over the past 25 years. These cover many diverse subjects and have titles such as "Most Women Who Maintain Families Receive Poor Job Market Returns," "One-Fourth of the Adult Labor Force Are College Graduates," and "Have Employment Patterns in Recessions Changed?"

THE CURRENT EMPLOYMENT STATISTICS PROGRAM

Given the amount of information available from the CPS, what is left for the Current Employment Statistics (CES) program to measure? The CES provides a vast array of information on employment, hours, and earnings by detailed industry and geographic location and is an important building block in national income estimates, the industrial production index, and the leading economic indicators. (See the essay by Moore on economic indicators.)

In contrast to the CPS, which is a survey of households, the CES is a survey of establishments. Data for the CES are collected from business payrolls by the Bureau of Labor Statistics in cooperation with state agencies. Much larger than the household survey, the CES sample includes almost 300,000 establishments employing more than 38 million people. Indeed, the CES sample is the largest monthly sampling operation in the field of social and economic statistics.

To collect CES data, state agencies mail forms to the sample establishments and review the returns. The states use the data to prepare state and area employment estimates and also transmit the reported data to BLS in Washington, D.C., for use in preparing national statistics. Each month, the same form is shuttled back to the establishments so that the next month's data can be entered below those from previous months. Firms report the total number of workers and the number of women in their establishments along with employment, payroll, and work hours of production or nonsupervisory employees.

The CES asks about employees on the payroll during the *pay period* that includes the 12th of the month, much like the household survey's reference to the *week* that includes the 12th. But the CES and the CPS do differ somewhat in their concepts. The establishment data exclude workers in agriculture, as well as those who are self-employed. Therefore the CES job count is smaller than "total employment" from the CPS.

The payroll data are tabulated by industry. Goods-producing industries are comprised of manufacturing, mining, and construction. The service-producing industries include transportation and public utilities; wholesale trade; retail trade; finance, insurance, and real estate; services; and government.

The scope of the establishment survey allows the Bureau of Labor Statistics to estimate employment for very detailed industries. For example, under the durable-goods manufacturing heading we find primary-metal industries, which are subdivided into five more detailed industries. Even these components are further divided to the point of differentiating between gray-iron foundries and malleable-iron foundries. We also learn how many workers in these foundries are production workers, their average weekly hours, overtime, and hourly and weekly earnings.

Why are two such complex surveys used to study the labor market? Simple. The strengths of one survey are generally the other's weaknesses. For example, the CES provides data on job trends in hundreds of detailed industries, but says little about the characteristics of the workers and nothing about people who are not working. The CPS, in contrast, is less detailed regarding industries, but is full of details about people, whether employed, unemployed, or not in the labor force.

WHAT SEASONAL ADJUSTMENT IS ALL ABOUT

Let's say you heard that employment—as measured by both surveys—declined last June. How could this be? Even if the economy were weak, you know that more people usually work in June than in May. After all, students out of school for the summer and new graduates flood the labor force and a good proportion of them find jobs. The unexpected news is the result of a procedure called *seasonal adjustment.* Seasonal variation in employment can be very large. For many data series, seasonality accounts for the bulk of the month-to-month changes. For example, construction employment always drops during the cold winter months, while retail trade jobs pick up during the Christmas season. The labor force expands each June, and agricultural employment grows at harvest times. Seasonal adjustment, essentially an averaging technique based on past experience, removes these seasonal changes from the data. Employment levels declining between May and June, "after seasonal adjustment," often indicate that the sharp increase that typically occurs during that period was smaller than usual. Data changes after seasonal adjustment are true economic changes, not simply those that occur each year at about the same time.

Without seasonal adjustment, it is difficult to determine the underlying trends in a series. Downturns or upswings in the business cycle are more readily spotted after seasonal adjustment, and long-term secular trends—such as the growth in employment over time—are more apparent.

Information from the government's two major employment surveys makes headlines, spurs debates, and excites scholars, politicians, and the business community. How they use the data is entirely up to them, but making sure that the data are timely and accurate is the job of the government. For without these surveys, our nation would know much less about how well or how poorly its economy is functioning.

PROBLEMS

1. Explain why the Census interviewer asks a series of questions about an individual's labor force activity rather than just asking "Are you unemployed?"
2. Why aren't the same households interviewed each month? Why aren't all new households interviewed each month?
3. Is unemployment bad? How can the "official" definition of unemployment be made broader or narrower to result in either a higher or lower jobless rate?
4. Why are many labor force indicators expressed as ratios, such as the unemployment rate or the employment-to-population ratio?
5. Why can't the government get a complete count of the unemployed from administrative records of the unemployment compensation system or by asking employers?
6. Why does the establishment survey yield more accurate estimates than does the household survey for specific industries?

7. In what ways do the employment estimates from the establishment and household surveys differ?

8. Why are employment and unemployment data seasonally adjusted? Why aren't the unadjusted data used to analyze over-the-month changes?

9. Is the figure reported for unemployment among black teenagers just as accurate, more accurate, or less accurate than that for total unemployment? Why?

10. Why does the Bureau of Labor Statistics call the overall unemployment rate "about unchanged" when it drops from 7.0% to 6.9%?

REFERENCES

U.S. Department of Labor, Bureau of Labor Statistics, April 1988. *BLS Handbook of Methods.* Bulletin 2285. Washington, D.C.: U.S. Government Printing Office.

U.S. Department of Labor, Bureau of Labor Statistics, September 1987. *How the Government Measures Unemployment.* Report 742. Washington, D.C.: U.S. Government Printing Office.

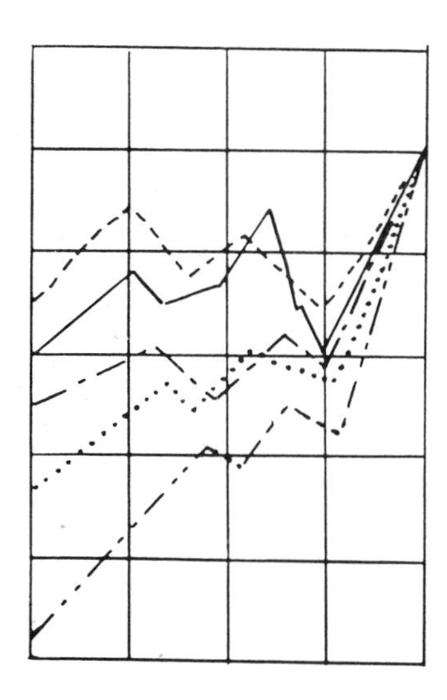

The Development and Analysis of Economic Indicators*

Geoffrey H. Moore *Center for International Business Cycle Research, Graduate School of Business, Columbia University*

INDICATORS AND BUSINESS CYCLES

Business cycles, large and small, appear to be a continuing feature of the economic landscape. A turn up or down in the economy is clearly an event of major social significance. Considerable interest therefore attaches to the means whereby an economic turn can be forecast and its extent can be estimated. That is the role of *economic indicators,* which rest on the numerous measurements of the pulse of the economy made by government agencies, private organizations, and individual economists. The analysis of economic indicators is a well-developed technique for ascertaining what the many pulse readings are saying about the state of the economy.

*This essay is adapted and updated from the author's article, "The Analysis of Economic Indicators," published in *Scientific American,* 232:1 (January 1975) (copyright 1974 by Scientific American, Inc., all rights reserved).

Economic indicators have come to embrace virtually all the quantitative measures of economic change that are continuously available. One can find daily, weekly, monthly, and quarterly indicators; they measure production, prices, incomes, employment, investment, inventories, sales, and so on; and they record plans, commitments, and anticipations as well as recent transactions. Some of the indicators, such as the unemployment rate or the Consumer Price Index, are calculated by the Federal Government on the basis of elaborate sampling surveys conducted each month. Others, such as the indexes of stock market prices and the surveys of purchasing agents' views of prices, orders, and inventories, are constructed by private organizations on the basis of information they collect or obtain as a by-product.

As a result economists or business people interested in forecasting change are faced, like weather forecasters, with a mass of factual information that pours in constantly. They must assess in some systematic way what the information says about the present and the future. The technique of *indicator analysis* embraces various systematic ways of looking at this information with a view to assessing the present situation and discerning significant future developments in the business cycle.

One of the earliest systems of the kind, devised shortly before World War I, came to be known as the Harvard ABC curves. The A curve was an index representing speculation, more specifically stock prices. The B curve represented business activity, measured by the dollar volume of checks drawn on bank deposits. The C curve represented the money market, measured by the rate of interest on short-term commercial loans. Historical studies, particularly those carried out by Warren Persons at Harvard University, showed that these three curves typically moved in sequence: stock prices first, bank debits next, and interest rates last, with the lagging turns in interest rates preceding the opposite turns in stock prices. The economic logic of the sequence was that tight money and high interest rates led to a decline in business prospects and a drop in stock prices, which led to cutbacks in investment and a recession in business. The recession in turn led to easier money and lower interest rates, which eventually improved business prospects, lifted stock prices, and generated a new expansion of economic activity.

The system came to grief in the Great Depression of 1929 because the interpreters of the curves took too optimistic a view and failed to foresee the debacle. Economists generally regard the episode as one of the great forecasting failures of all time.

Curiously, however, the sequence of events on which the system was originally based has in large measure persisted. This is not to say that the ABC curves would still suffice if they were revived. Far more comprehensive systems of indicators have been developed since 1929, and the empirical and theoretical base on which they stand has been more thoroughly studied, documented, and tested.

The sharp recession of 1937–1938, which occurred before the economy had fully recovered from the Great Depression, helped to spur that development. In the autumn of 1937, Henry Morgenthau, Jr., the Secretary of the Treasury, asked the National Bureau of Economic Research (a private, nonprofit research agency) to devise a system of indicators that would signal when the recession was nearing an end. At that time the quantitative analysis of economic performance in the United States did not approach today's standards. The Government's national income and product accounts, which form the foundation of much of modern economic analysis, were just being established. Other vital economic statistics, including unemployment rates, were being developed or refined by public agencies trying to provide information that would be useful in fighting the Depression. Few statistical series were issued in seasonally adjusted form, as they are now. Comprehensive econometric models (systems of equations expressing quantitative relations among economic variables), which are widely employed now to forecast the economy and to evaluate economic policies, were virtually unknown then.

Under the leadership of Wesley C. Mitchell and Arthur F. Burns, the National Bureau of Economic Research had since the 1920s assembled and analyzed monthly, quarterly, and annual data on prices, employment, production, and other factors as part of a major research effort aimed at gaining a better understanding of business cycles. This project enabled Mitchell and Burns to select a number of series that, on the basis of past performance and of relevance in the business cycle, promised to be fairly reliable indicators of business revival. The list was given to the Treasury Department late in 1937 in response to Morgenthau's request and was published in May 1938. Thus originated the system of leading, coincident, and lagging indicators widely employed today in analyzing the economic situation, determining what factors are favorable or unfavorable, and forecasting short-term developments.

Since 1938 the availability and the use of economic indicators have been greatly expanded under the leadership of the National Bureau of Economic Research, the U.S. Department of Commerce, the Organization for Economic Cooperation and Development in Paris, and other public and private agencies. The list of indicators assembled in 1937 was revised in 1950, 1960, 1966, and 1975 to take into account new economic series, new research findings, and changes in the structure of the economy. A new evaluation has recently been conducted by the Center for International Business Cycle Research at Columbia University. With each revision the performance of the indicators both before and after the date of their selection has been carefully examined and exposed to public scrutiny.

In 1957 Raymond J. Saulnier, who was then chairman of the President's Council of Economic Advisers, asked the Bureau of the Census to develop methods whereby the appraisal of current business fluctuations could take advantage of the large-scale electronic data processing that was becoming available, with the results to be issued in a monthly report. Experimental work done over the next few years under the leadership of Julius Shiskin, who was the chief economic statistician of the Bureau of the Census, resulted (in 1961) in the monthly publication by the Department of Commerce of *Business Cycle Developments*. (It is now called *Business Conditions Digest*; under both names economists refer to it as BCD.) This publication has greatly increased the accessibility of current indicator data and of various statistical devices that aid in their interpretation. As a result the indicators have become a major economic forecasting tool.

USEFUL QUALITIES OF INDICATORS

As noted above, the analysis of economic indicators rests on both an empirical and a theoretical foundation. The selection of particular indicators and the emphasis given to them have been guided by what is understood of the causes of business cycles. Obviously one would wish to examine recent changes in any economic process that is believed to play a significant role in any widely accepted explanation of cyclical fluctuations.

Many different explanations have been advanced for these fluctuations. Some of them place primary stress on the swings in investment in inventory and new plant and equipment that both determine and are determined by movements in final demand. Others assign a central role to the supply of money and credit, or to Government spending and tax policies, or to relations among prices, costs, and profits.

All these factors undoubtedly influence the course of business activity. Some of them may be more important at a given time than others. No consensus exists, however, on which is the most important or even on how they all interact. Hence it is prudent to work with a variety of indicators representing a broad range of influences. Ready access to a wide range of indicator data enables one to test competing or complementary hypotheses about current economic fluctuations.

With this principle in mind, economic activities can be classified into a few broad categories of closely related processes that are significant from the business cycle point of view. Indicators have been selected from each group. The principal categories now included in *Business Conditions Digest* are shown in the lefthand column of Table 1. Note that these categories do not include all aspects of the economy. For example, statistics on agriculture; Federal, state, and local government; foreign trade; and population and wealth are omitted. Nevertheless, the categories do provide a framework of factors that enter into theories of the business cycle and are important in assessing the performance of the economy. The omitted categories are important, too, but supply few indicators that are systematically related to the business cycle.

Within each category, research on business cycles has uncovered indicators that behave in a systematic way. These findings have provided a basis for selecting particular indicators and classifying them according to their characteristic cyclical behavior, as in Table 1. Two of the chief characteristics one looks for are the regularity with which the indicator conforms to business cycles and the consistency with which it leads, coincides, or lags at turning points in the cycles. Other relevant considerations are the statistical adequacy of the data (since the statistical underpinning of an indicator has a bearing on how well the indicator represents the process it is supposed to reflect), the smoothness

Table 1 Cross-classification of economic indicators

	Relation to Business Cycle		
Economic Process	*Leading*	*Roughly Coincident*	*Lagging*
Employment and unemployment	Average workweek and overtime Hiring and layoff rates New unemployment insurance claims	Total employment	Long-duration unemployment
Production and income		Real GNP Industrial production Personal income	
Consumption, trade, orders, and deliveries	New orders, consumer goods Vendor performance	Retail sales Manufacturing and trade sales	
Fixed capital investment	New investment commitments Formation of business enterprises Residential construction		Backlog of investment commitments Investment expenditures
Inventories and inventory investment	Inventory investment and purchasing		Inventory levels
Prices, costs, and profits	Change in industrial materials prices Stock prices Profits and profit margins	Change in producer prices	Change in consumer prices Change in unit labor costs
Money and credit	Money and credit flows Credit delinquencies and business failures Bond prices		Outstanding debt Interest rates

of the data (since highly erratic series are difficult to interpret correctly), and the promptness with which the figures are published (since out-of-date figures have a limited bearing on the current situation).

Empirical measures of these characteristics have been constructed for large numbers of indicators. Such measures have been employed in the attempt to obtain data capable of conveying an adequate picture of the changes in the economy as it moves through stages of prosperity and recession. In addition, the behavior of the indicators after they have been selected has been monitored closely. Many of the indicators have survived several successive evaluations. For example, measures of the average workweek, construction contracts, and stock prices have been on every one of the successive lists of indicators that have been drawn up by the National Bureau of Economic Research since 1937.

The same lists of indicators have also been tested by their performance in other countries. Every new recession or slowdown, whether in this country or abroad, provides additional evidence against which the indicators can be assessed, as does every upturn. This examination and reexamination has accumulated a large amount of empirical evidence that demonstrates both the value of the indicators and their limitations.

SEASONAL ADJUSTMENT AND SMOOTHING

A sampling of this evidence is contained in the accompanying illustrations. Let us consider an indicator such as the number of building permits issued for new houses (see Figure 1). The raw data are statistically decomposed in order to measure and eliminate regular seasonal variations that are repeated every year. When this factor is removed, the indicator reveals much more clearly the tendency for permits to diminish during the recessions of 1980 and 1981–1982 and to increase when more prosperous conditions returned in 1983.

Most economic indicators today are available in seasonally adjusted form. Some of them are seldom reported in any other way. Examples of indicators that are invariably adjusted for seasonal factors include the gross national product, the unemployment rate, and the index of industrial production. One of the computer programs that is widely used to make seasonal adjustments, developed by the Census Bureau, is called the X-11 program. (See also the article by Leon and Rones for a discussion of seasonal adjustment.)

The smoothing of irregular movements is less commonly practiced because the techniques are somewhat less routine. Certain statistical series, however, are subject to much wider irregular movements than others because of differences in sampling error or in the effects of such factors as unusual weather or labor disputes. It is therefore useful in interpreting current changes to have a standard measure of the size of these irregular fluctuations compared with the size of the movements that reflect long-term trends and the events of the business cycle, which are often called *trend-cycle movements.*

Two measures of this kind are provided for all the indicators carried in *Business Conditions Digest.* One shows how large the average monthly change

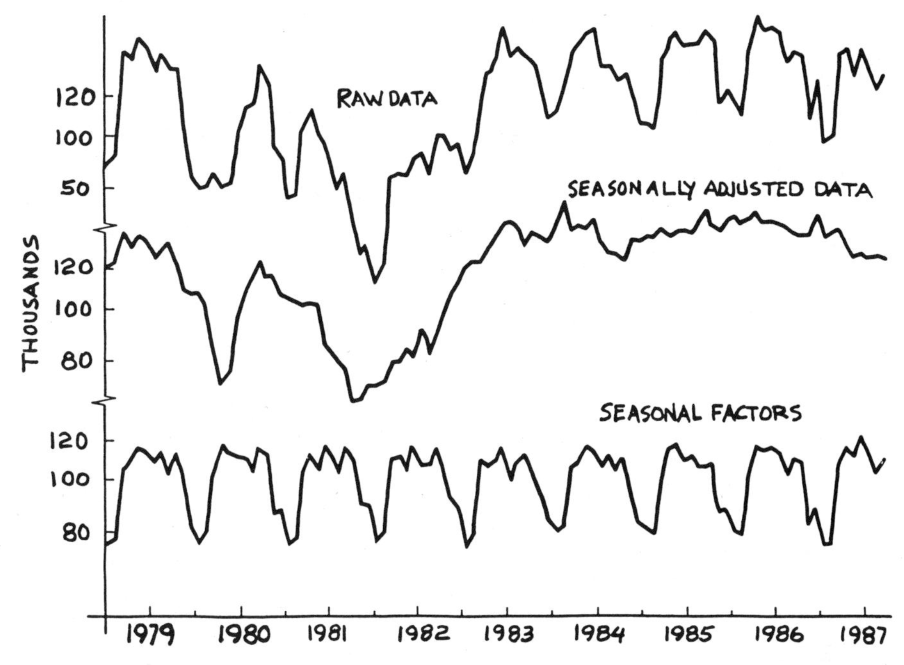

Figure 1 *Building permits for new houses, before and after seasonal adjustment. Source: Department of Commerce.*

in the irregular component is with respect to the average monthly change in the trend-cycle component. The other shows how many months must elapse on the average before the change in the trend-cycle component, which builds up over a period of time, exceeds the irregular component, which does not. For example, the measures show that monthly changes in housing starts are likely to be dominated by "noise" (such as random variation), but that when these changes are measured over spans of four months, the trend-cycle "signal" becomes dominant. On the other hand, the index of industrial production is much less affected by noise, so that monthly movements are more significant.

LEADS AND LAGS

The most important characteristic of an indicator from the point of view of forecasting is of course the evidence it provides concerning future changes in economic activity. Indicators differ in this respect for numerous reasons. Certain types, such as housing starts, contracts for construction, and new orders for machinery and equipment, represent an early stage in the process of making decisions on investment. Since it takes time to build a house or a factory or a turbine, the actual production (or completion or shipping) usually lags behind the orders or contracts. The lag depends on, among other things, the volume of unfilled orders or contracts still to be completed. Where goods are made for stock rather than to order there may be no lag because orders are filled as they are received.

Another kind of lead-lag relation exists between changes in the workweek on the one hand and employment on the other. In many enterprises employers can increase or decrease hours of work more quickly, more cheaply, and with less of a commitment than they can hire or fire workers. Hence in most manufacturing industries the average length of the workweek usually begins to increase or decrease before a corresponding change in the level of employment. The workweek is therefore a leading indicator with respect to the unemployment rate.

Many bilateral relations of this kind have been traced (see Figure 2). The matter obviously becomes more complex, however, when the relations are multilateral. Indexes of stock market prices, for example, have exhibited a longstanding tendency to lead changes in business activities (the Harvard ABC curves relied in part on this tendency), but the explanation seems to require the interaction of movements in profits and in interest rates, and other factors as well. A cyclical decline in profits often starts before a business expansion comes to an end; the proximate cause is usually a rapid rise in the costs of production. Interest rates also are likely to rise sharply. Both factors operate to reduce the attractiveness of common stocks and depress their prices, even though the volume of business activity is still rising. Near the end of a recession the opposite tendencies come into play and lift stock prices before business begins to improve. Since interest rates are generally counted among the lagging indicators, this case illustrates that even lagging indicators can play a role in forecasting, by affecting the movements of the leading indicators.

For the purpose of measuring leads and lags a chronology of business cycles has proved useful. The National Bureau of Economic Research has defined business cycles in such a way that peaks and troughs can be dated with reasonable objectivity. Indeed, some parts of the dating procedure can now be carried out by computer. Since the vast majority of indicators that are of interest show cyclical movements conforming to business cycles, the peaks and troughs in each indicator can be matched with those of the business cycle to determine characteristic leads and lags.

Following this plan, groups of indicator series that typically lead, coincide with, or lag behind turns in the business cycle have been identified, as illustrated in Table 1. Composite indexes constructed from these groups (see Figure 3) can be employed (as individual indicators can) to measure the relative severity of an economic downturn as it progresses from month to month. With such a monitoring scheme one can observe the relative severity of the current decline and draw certain inferences based on the fact that the severity rankings among different recessions have usually not changed a great deal after the first few months.

It is of course essential in any appraisal of the economic outlook to take into account actual and prospective policy actions by the Government. Such actions include tax reductions or increases, changes in required bank reserves, changes in military expenditures, and the establishment of programs of public employment. Such actions often do not fit readily into the framework of indicators,

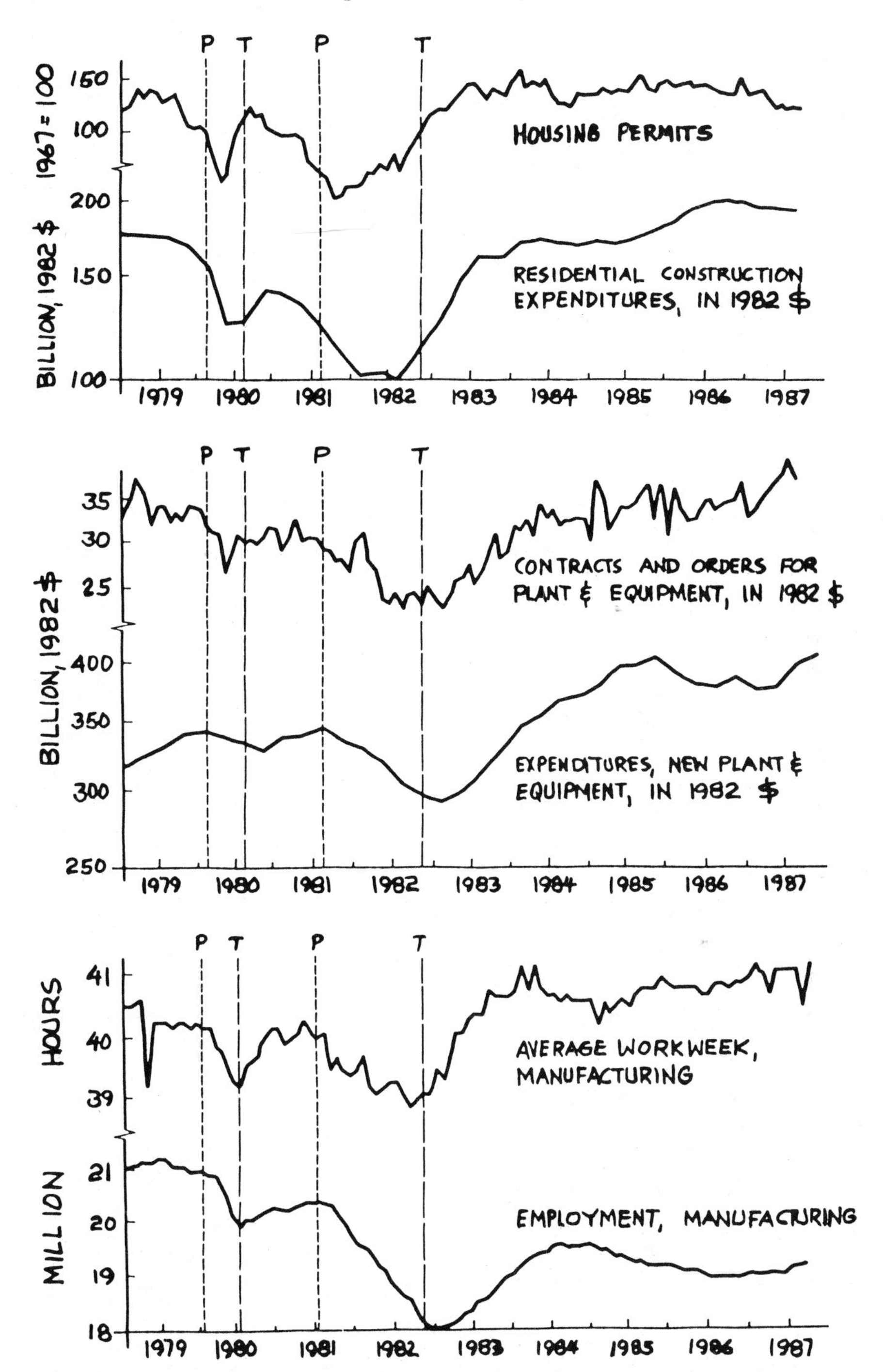

Figure 2 *Three leading indicators and the activities they lead. Note: Vertical lines are business cycle peaks (P) and troughs (T). Sources: Department of Commerce and Bureau of Labor Statistics*

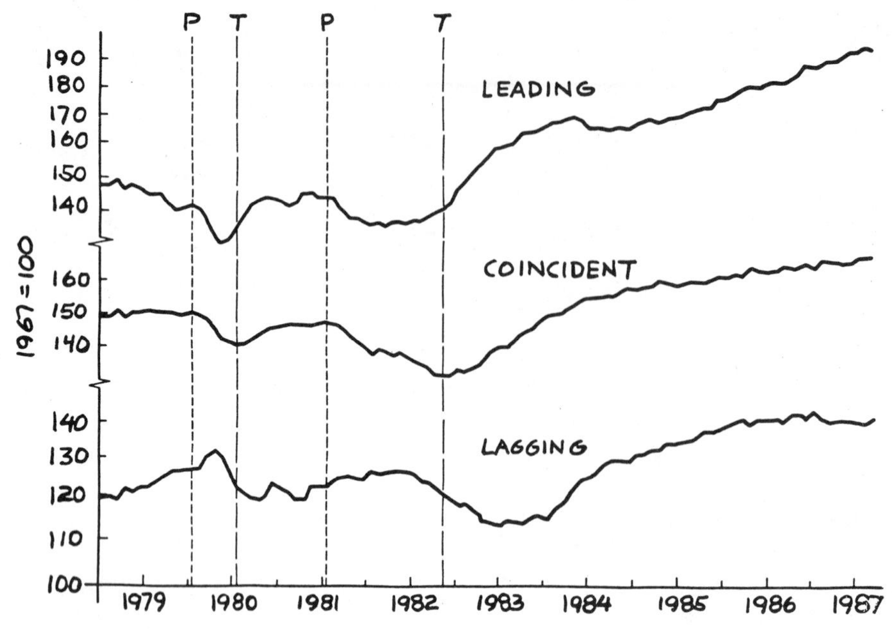

Figure 3 *Leading, coincident, and lagging indexes for the United States. Note: Vertical lines represent business cycle peaks (P) and troughs (T). Source: Department of Commerce.*

although their effects, together with other influences, may be registered promptly in orders, contracts, housing starts, stock prices, and so on. Still, certain indicators do provide a nearly continuous reading on Government activities, although they have generally not been found sufficiently systematic in their relation to business cycles to warrant selection as leading, coincident, or lagging indicators.

INTERNATIONAL ASPECTS

Economic indicators of all types are followed more widely in the United States than in most other countries. The growth of trade, travel, and international finance, however, has increased the need for promptly available statistics on international transactions and economic development in other countries. In 1973 the National Bureau of Economic Research began a program of assembling and analyzing indicators for a dozen industrial countries, and since 1979 this program has been carried forward by the Center for International Business Cycle Research.

Fortunately the indicator approach is sufficiently flexible to be adapted readily to situations where, as in many countries since World War II, economic recessions have taken the form of retarded growth of aggregate activity rather than absolute declines and where such retardation may have been deliberately induced by government policies in order to cool off inflation or restore a deteriorating trade balance. Moreover, the approach is flexible enough to accommodate differences among countries in the types of indicator data that are available or are most revealing. For example, in Europe, statistics on job vacancies are relied on more than in the United States, and data on the international migration of workers are more significant because migrant workers are a significant part of the work force.

The pursuit of economic indicator analysis on an international scale, by international agencies as well as by domestic institutions, has demonstrated the feasibility of the approach and its potential value in observing and appraising international fluctuations in economic growth rates and the accompanying trends in price levels, foreign trade, capital investment, and employment. One can envision the evolution of a worldwide system of indicators, built on the plan originally developed for the United States, to support the analysis of economic indicators on a global scale.

PROBLEMS

1. Why are seasonally adjusted data used in analyzing business cycles?
2. What do the leading indicators lead?
3. What qualifications should a good economic indicator possess?
4. Political candidate A says a recession has begun because stock prices have dropped. B says no, because employment is still rising. How would you decide who is right?
5. The following figures are from the July 1975 issue of *Business Conditions Digest.*

	Leading Index	*Coincident Index (1967 = 100)*	*Lagging Index*
July 1974	145.3	138.8	210.5
Aug. 1974	140.4	138.2	214.5
Sept. 1974	135.0	137.4	216.3
Oct. 1974	130.1	136.2	219.0
Nov. 1974	126.0	132.3	220.4
Dec. 1974	123.6	128.2	220.0
Jan. 1975	118.7	125.2	217.8
Feb. 1975	118.6	124.1	212.9
Mar. 1975	120.3	122.0	210.1
Apr. 1975	124.8	122.2	205.5
May 1975	127.9	122.4	201.5
June 1975	130.7	124.6	200.2

Considering the fact that a recession began in November 1973, do you think it has ended and if so, when? What confirming evidence would you like to have?

6. How can leading indicators for another country, say, Japan, be helpful to business firms in the United States?

7. In deciding whether to invest in stocks or in bonds, what economic indicators would you examine?

RELATED REFERENCES AND DATA SOURCES

Bureau of Economic Analysis, U.S. Department of Commerce, 1984. *Handbook of Cyclical Indicators.* Washington, D.C.

Bureau of Economic Analysis, U. S. Department of Commerce. *Business Conditions Digest,* monthly. Washington, D.C.

Philip A. Klein and Geoffrey H. Moore. 1985. *Monitoring Growth Cycles in Market-Oriented Countries.* Cambridge, MA: Ballinger Publishing.

Geoffrey H. Moore. 1983. *Business Cycles, Inflation, and Forecasting,* 2nd edition. Cambridge, MA: Ballinger Publishing.

Organization for Economic Cooperation and Development. *Main Economic Indicators,* monthly. Paris.

PART FOUR
Our Physical World

Optimization and the Traveling Salesman Problem

Charles A. Whitney *Harvard University*

INTRODUCTION

An astronomer at the controls of a telescope often wishes to make relatively short measurements of a long list of stars scattered across the sky. Moving the telescope from one star to another is a time-consuming process, so the astronomer tries to select the most efficient sequence of "visitations," avoiding unnecessary motion. This problem is particularly acute with a space telescope orbiting the Earth high above the atmosphere. The pointing of such a telescope is often controlled from the ground by small gas jets and the available fuel is severely limited. If the stars are haphazardly distributed around the sky, the sequence must be carefully chosen to conserve time and fuel.

This is a form of the "traveling salesman problem," and it is usually expressed as the search for the shortest closed route among a set of cities such that each city is visited just once. It is typical of a class of problems that can be stated briefly but that defy direct solution and can only be solved by a series of educated guesses. These guesses are often based on probabilistic calculations. This property distinguishes such problems from most of the problems described in this book, and it describes many practical problems in fields as different as geological exploration and electronics.

A direct approach to the solution would be to locate the cities on a map, measure all the possible routes, and select the shortest. But that is easier said than done, as we can see by counting the routes. If there are N cities and the starting point is prescribed, there will be $N - 1$ choices for the first stop, $N - 2$ for the second, and so forth, making a total of $(N - 1) \times (N - 2) \times (N - 3) \ldots 2 \times 1$ possible routes back to the starting point. Even with as few as 8 cities, this leads to $7 \times 6 \times 5 \times 4 \times 3 \times 2 \times 1 = 5{,}040$ possible routes to be examined. And the number increases rapidly with the number of cities. If we have two groups of 8 cities, making a total of 16, there would be more than a trillion routes! And 16 is not a particularly large number, so we evidently must give up the idea of a direct measurement of all possibilities.

Another example is the design of electronic circuits, where dozens of components are to be connected by wires. The designer needs a method of finding an efficient layout that will minimize the amount of wire required. In modern circuits the number of possible arrangements often exceeds a trillion, and the designer would not have time in the life of the universe to try all possible solutions while looking for the best.

APPROACHING A SOLUTION OF THE TRAVELING SALESMAN PROBLEM

There is an aspect of this type of problem that gives us some hope of coming to a practical solution. We know intuitively that, in addition to the best solution, there will be many solutions that are nearly as good. Thus many of the good solutions, including the best, will be of nearly the same length. This means that we can be satisfied with a technique that brings us close to the optimum without worrying too much if we do not find the single best route. For problems of this type we can often assume that second best—or thousandth best—is good enough.

What more can we say about the possible solutions to the traveling salesman problem? Our intuition suggests—and experience verifies—that the traveler ought to complete the tour of each section of the map before moving to another section. As an extreme illustration, suppose the cities are arranged in two groups ($N/2$ in each group) separated by a distance of G miles, as illustrated in Figure 1. A route that jumps back and forth between the two groups would clearly not be an efficient choice (see Figure 2). Its length would be roughly equal to the product, $N \times G$, and the salesman would do better to finish one group of $N/2$ cities before moving to the other, as in Figure 3. For example, suppose D is the average distance between cities in each group, and the salesman starts from a selected city in the first group. A visit to all the remaining cities in the first

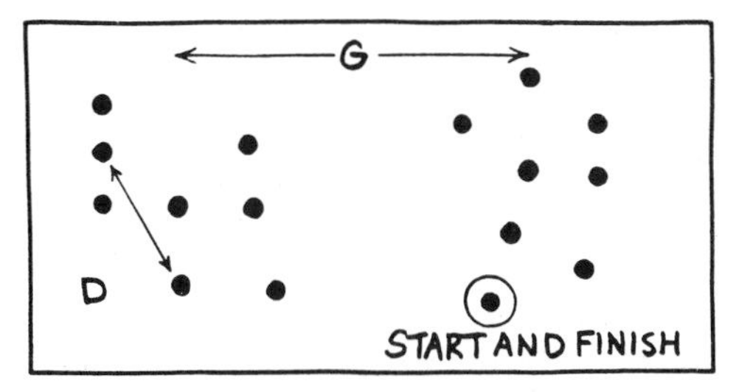

Figure 1 *Schematic map of 16 cities. The traveling salesman problem is to find the shortest route from a specified city (circled) that passes just once through all the other cities. In this example the cities are divided into two groups separated by a distance of G miles. Within each group the mean separation is D miles.*

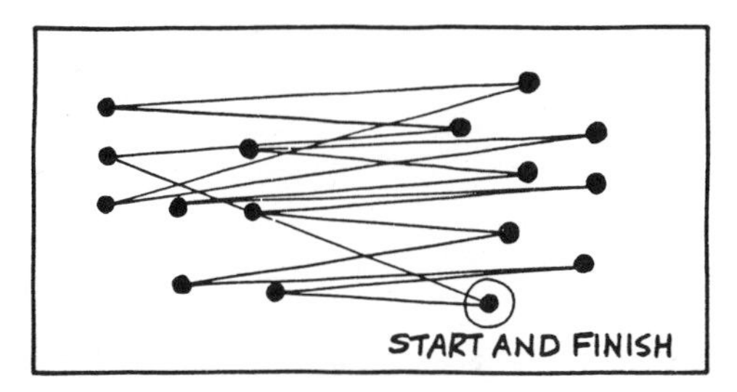

Figure 2 *Example of a poor solution, in which the salesman moves back and forth between the two groups of cities. His route length is approximately $N \times G$ miles, where N is the total number of cities. Figure 3 shows a better solution.*

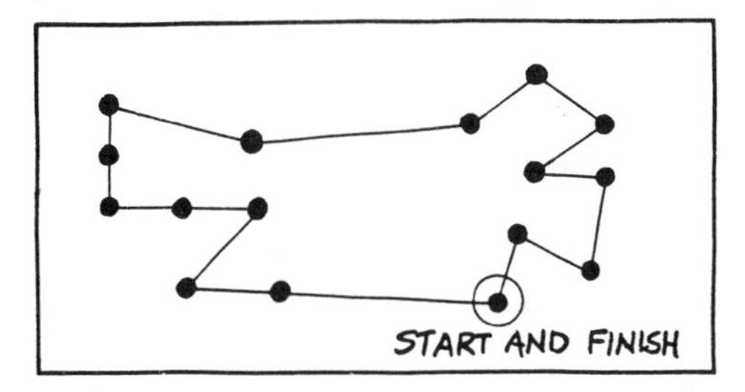

Figure 3 *Example of a good solution, in which the salesman visits all the cities in one group before moving to the other. As shown in the text, his route length in this case is approximately $L' = 2 \times G + D \times (N - 2)$, which is less than the distance $L = N \times G$, for the route in Figure 2. As an example, suppose $N = 16$, $G = 8$ miles, and $D = 2$ miles. Then $L = 128$ miles for the route in Figure 2, and $L' = 2 \times 8 + 2 \times 14 = 44$ miles. Thus the length of the better route is about ⅓ that of the poorer route.*

group would involve a distance $D \times (N/2 - 1)$. Then he would move G miles to a city in the other group, followed by $D \times (N/2 - 1)$ miles in visiting the remainder of the group. Finally, he would move G miles to get back to the first group, for a total of $2G + D \times (N - 2)$ miles. Figures 2 and 3 compare the lengths of two such solutions, and they show that a little care can lead to a substantial shortening of the route.

So the salesman ought to cover each region thoroughly before moving to the next, but how does he search for a good route within each region?

Before tackling this problem, let's consider a more general type of problem that has many industrial and scientific applications. It may appear to be a new type, but it is simply a variation on the traveling salesman problem.

Consider the cost of fuel for the operation of a simplified manufacturing plant shown schematically in Figure 4. The plant buys electricity from a public utility to operate electric equipment such as power saws and an air conditioner. The plant also buys fuel oil to operate a generator that produces electricity and a certain amount of heat that can also be used for controlling the air temperature. Given the costs of electricity and oil, as well as the efficiency of the machines—which depends on the amount of effort put out by each machine—how should the plant manager allocate the money available for fuel, and how should the various types of fuel best be used within the plant? And if the plant needs more air conditioning capacity, what type of new machine ought to be brought in? In a realistically complex problem, the solution is far from obvious.

GENERAL DESCRIPTION OF THE PROBLEM OF OPTIMIZATION

All of these problems have several features in common. First, for each, we can specify a *cost function* whose value we seek to reduce to a minimum. It may be the length of a route, or the cost of fuel, or the total time required to point a space telescope at a set of selected stars. Second, each problem is specified by a number of fixed parameters that are outside our control, such as the positions of the stars in the sky, or the fuel requirements of the factory machines and the manufacturing quotas set by management. These are *constraints* on the problem, and they limit the possible solutions. Third, there are the adjustable numbers whose choice constitutes the solution of the problem. For example, we adjust the sequence of stars until the time is minimized, or we adjust the purchases of fuel and the allocation of machine output in the factory to minimize the overall cost of operation.

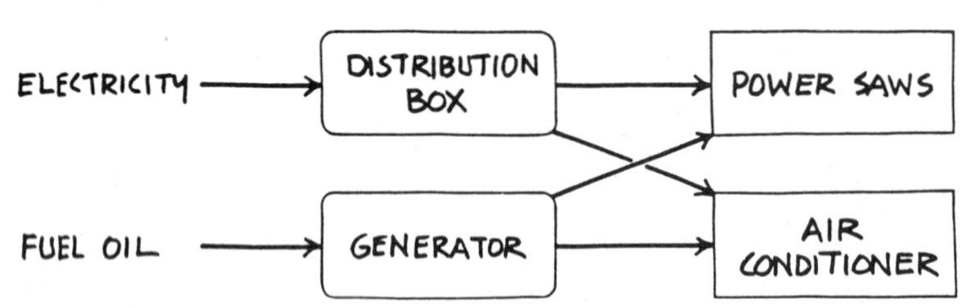

Figure 4 *This diagram of a manufacturing plant that operates power saws and an air conditioner provides a simplified industrial example of the problems described in this essay. Electricity and gas have different costs and efficiencies, and the plant manager must decide on the most economical amounts of electricity and oil to buy. The answer will depend on deciding how much electricity to produce with the plant's own generator. In a real plant, the problem is often too complicated to be solved directly, and the manager must resort to random search techniques.*

With the advent of relatively inexpensive electronic computers, it has become feasible to carry out such searches by groping in the dark, so to speak. This process is very much like a randomized search for the deepest part of a lake.

To take a problem that arises in geological exploration, suppose we are prospecting for gas and are probing for the top of an underground gas dome, trapped above an oil field (Figure 5a). In order to get the largest quantity of gas, we wish to tap it at the highest part of the dome. Assume we have a device for measuring the height of the dome above the level surface of the oil at any point, and we look for the position that makes the height a maximum. For consistent terminology, we wish to express the problem in terms of minimizing a cost function, so let us define the depth of the dome below a convenient level surface as the cost function and seek to minimize it.

If we knew very little about the dome we would start at an arbitrary place. We would read the depth at the starting point, then move a short distance and again measure the depth. If it decreased, we would continue moving in that direction. If the depth increased, we would move in another direction. The selection of a new direction could be based on the local stratification or it might be a blind guess. In either case, we would continue this process until we came to a place where the depth increased in all directions. We would then know we were near the peak of the gas dome. All points within a small distance of the peak—say, a few meters—would have nearly the same depth because we can consider the top to be horizontal in the region of the peak. So we could adopt any of the positions in this region as the solution, as it would make no practical sense to insist on locating the peak to the nearest centimeter.

If the dome is actually a smooth spherical or cylindrical shape, we can be fairly sure we have come to the neighborhood of the true peak. But suppose the shape is more complex, as in Figure 5*b*. In that case, we may merely have found a localized peak. How can we avoid mistaking a local peak for the true top? The key is in a ***probabilistic*** approach that keeps us from getting caught at a false peak.

DESCRIPTION OF A PROBABILISTIC SEARCH FOR THE OPTIMUM

In a probabilistic search we proceed by successive corrections, starting from a guessed solution and trying new, randomly generated solutions until we are satisfied. To see how this might work, let C be the cost function to be minimized, say, the length of the route or the local depth of the gas dome, for a particular solution. We evaluate C for the first guess and then construct a new solution—randomly or using whatever information we have. This is equivalent to moving to another position over the gas dome. We then reevaluate the cost function (depth), obtaining C'. If the new value is less than or equal to the previous, $C' \leq C$, we accept the new position as defining a next starting point. On the other hand, if the new depth is greater, $C' > C$, we decide whether to reject or accept the new position on the basis of a probabilistic calculation, as follows. We compute a random number (imitating the outcome of tossing a pair of dice, for example), and we restart from the old position if this number

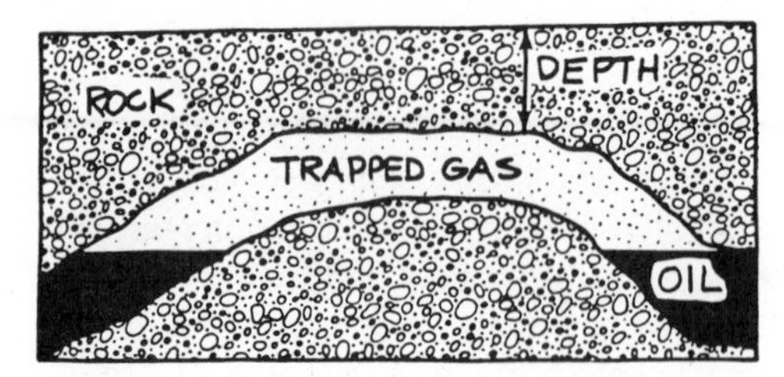

Figure 5a *Diagram of a hypothetical oil deposit with an overlying region of natural gas. The depth of the gas dome below the surface of the ground is indicated. When the gas is tapped off, the oil will rise in the dome, so the tap is to be placed at the highest peak of the gas dome to get all of the gas. The text describes a random search for the peak, where the depth is a minimum.*

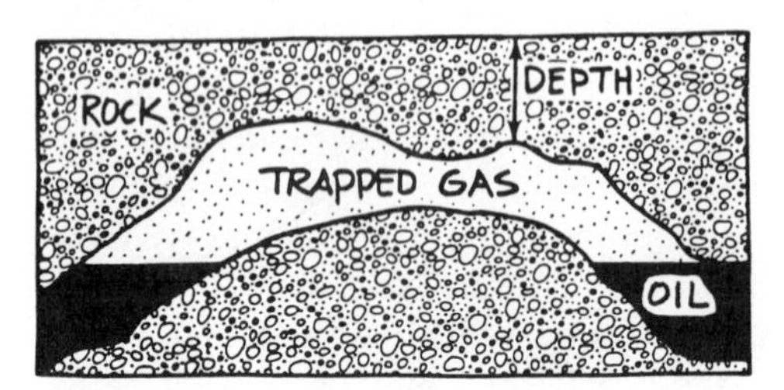

Figure 5b *Similar to Figure 5a, except the dome has a secondary peak that might be mistaken for the true peak. The search technique must be designed to avoid stopping at the secondary peak, or the gas in the higher peak will be lost.*

is less than some other number we have chosen previously. We repeat this process many times, moving from one point to the next.

Let us not worry, for the moment, about the details of the calculation that leads us to accept or reject a new solution because the details are not crucial to an understanding of the method. The essential point is that the probabilistic decision permits us to occasionally move to a new point even if it entails a small *increase* of *C*. This prevents our getting stuck in local minima before we reach the neighborhood of the absolute minimum. This is the way we avoid stopping at a secondary peak of the gas dome. We always permit steps that would momentarily take us to a deeper point of the gas dome because sometimes that is the only path to the true peak.

Despite the fact that the process is rather like groping in the dark, it has great power for two reasons. First, it searches among a small sample of possible solutions and picks out solutions that are approximately the best. This greatly reduces the search time. Second, it can be used to find a practical solution to any problem that can be expressed in terms of a cost function, *C*, and a set of adjustable parameters. As long as a computer is available, we need not care whether the numerical evaluation of the cost function is simple or complex. The sky is the limit, and this explains the usefulness of the random search for problems as diverse as manufacturing, geological exploration, and astronomy.

For those of you who are interested, I'll describe some of the technical details of the probabilistic decision to accept or reject each new solution. I will phrase it in terms of the randomized search for the peak of a gas dome.

After each new position has been selected and its depth, C', has been evaluated, we divide the change of cost, $C' - C$ by a number, T, which has the same units as C. (The selection of the value of T is a matter of experience. It must be suited to the problem.) We next compute a random number, r, in the range between zero and unity, $0 \leq r < 1$. We take its natural logarithm (which will always be a negative number) and compare the result with $-(C' - C)/T$. If

$$\text{logarithm } r < -(C' - C)/T,$$

we accept the new route; otherwise we reject it.

This particular formula is chosen because it is very permissive about small increases of C while permitting only a very small number of large increases of C. The probability of accepting an increase of C depends on the value of T. If T is large, this formula permits large increases. This means that almost every new trial we make will be accepted as a new starting point. This is a good way to start a search because we can wander freely about the dome. As the computation progresses, it will sample the dome here and there and move among the various peaks. The rejection criterion is gradually adjusted by reducing the value of T, making the acceptance of an increase of C less likely. After all, a geologist who has already sampled much of the gas dome will not want to move back to where the dome is lower. Establishing the schedule for reducing T is a matter of experience; if T is reduced too quickly, the process may get stuck at a false peak of the gas dome; if it is reduced too slowly, the geologist will wander indefinitely. Ideally, the reduction of T will gradually restrict the search to the neighborhood of the true peak. The process is stopped when T has become small and the search appears confined to a small region of the dome. The geologist can confirm that the true peak has been found by repeating the entire search and seeing whether it returns to the same region.

The effect of changing T is illustrated in Figure 6*a* and *b*, where two hypothetical searches are traced. In the first case, which would be appropriate to the early stages of a search, the large value of T permits probing much of the dome. When the search wanders back into the vicinity of the true peak, the value of T is decreased, thus trapping the search in the best region.

CONCLUSION

The random search method, despite its apparent blindness, has proven very powerful. The key to its success is twofold. It samples a small number of the total number of possible solutions, and it avoids being trapped by false solutions. A good solution for virtually any problem that can be expressed in terms of a cost function and a set of adjustable numbers can be found by this method. How close the derived solution comes to the best possible solution depends on the persistence of the solver, but in most practical problems, a good solution is entirely adequate.

Figure 6*a* *Contour map of the roof of a gas dome with two peaks. The black region indicates the oil level; lighter shading indicates the higher roof, and the true peak is on the left. A random search with a large value of the parameter, T, is indicated by the segmented line. It covers a large portion of the dome, and it will not settle down in either peak until T is decreased.*

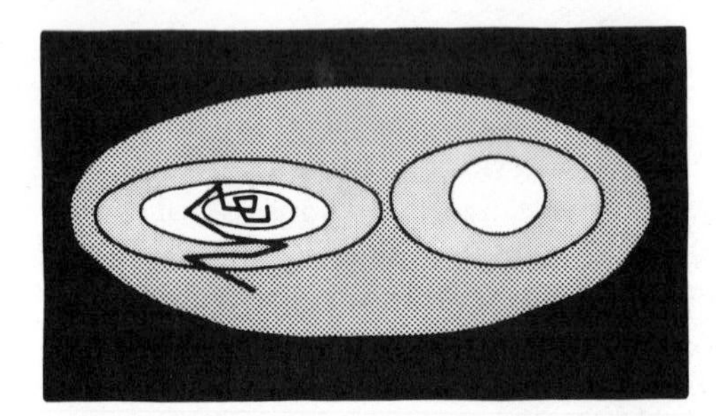

Figure 6*b* *Similar to Figure 6a. In this case, a small value of T was used and this confines the search to steps that move toward the nearest peak, so it can stop at the wrong peak, although in this case it did find the higher peak.*

PROBLEMS

1. List several types of problems that might be amenable to the random search method. What are the characteristics of a problem that would require this type of approach?

2. Imagine you have a newspaper route and wish to find the most efficient delivery sequence. What are some of the ways you might construct trial routes? Suppose you live in a suburban area where the houses are strung along several roads with few cross streets. How would this affect the process by which you would construct trial routes?

REFERENCE

S. Kirkpatrick, C. D. Gelatt, Jr., and M. P. Vecchi. 1983. "Optimization by Simulated Annealing." *Science* 220: 671–680.

Estimating the Chances of Large Earthquakes by Radiocarbon Dating and Statistical Modeling*

David R. Brillinger *University of California, Berkeley*

Residents in earthquake-prone areas are concerned with the possibility that an earthquake might occur and cause them loss of life or property. They seek insurance to reduce the effects of this risk. Government officials are also concerned, for they have the responsibilities of planning continuing services if there is damage to critical facilities and for educating the public. Basic to these insurance premium calculations and government allocation of resources are estimates of the chances of earthquakes and of the associated destruction.

Fortunately large earthquakes are rare. Unfortunately, however, their rarity has the statistical disadvantage of making it difficult to estimate their chances of occurrence confidently. Several procedures have been developed to assess seismic risk. This essay describes a cross-disciplinary approach that has the

*This research was partially supported by USGS Grant 14-08-0001-G1085. Helpful comments were received from Professors R. Purves, K. Sieh, M. Stuiver, and from the editors.

wonderful aspect of being based on data for earthquakes that occurred at a location of interest when no one was there to record the event. In fact, nine of the ten earthquakes employed in the study are prehistoric.

PALLETT CREEK

This story began with Stanford geologist Kerry Sieh heading off into the California desert just after his honeymoon, in the company of his wife and brother. Professor Sieh's destination was a small piece of ground straddling the San Andreas Fault about 55 kilometers northeast of Los Angeles (see Figure 1 for the general location). A stream called Pallett Creek runs nearby. Until 1910 or so this area was a swamp. Over the years, black peats were formed and periodically buried by sand and gravel borne by the creek's floodwaters. Sieh and his companions proceeded to dig trenches. They found disrupted layers of peat, wood fragments, charcoal, and even old animal burrows. Examining the trench walls Sieh noted places where the layers were broken and inferred that these breaks occurred during prehistoric earthquakes. All of his professional train-

Figure 1 *Map of California showing location of site on San Andreas Fault where specimens were collected for radiocarbon dating. PC indicates the location of Pallett Creek, LA indicates Los Angeles, and UCB the University of California, Berkeley.*

ing and expertise as a geologist helped him to decide which disruptions in the layers might correspond to earthquakes. He selected specimens near each of the breaks to date by radiocarbon techniques.

In the way of background, the most recent large earthquake that affected the Pallett Creek area was in 1857. (A large earthquake is one of Richter magnitude 7.5 or greater.) Also, the study of earthquakes at Pallett Creek is highly informative concerning destructive events that might hit the greater Los Angeles area.

RADIOCARBON ANALYSIS

After returning from Pallett Creek, Sieh sent the specimens to Professor Minze Stuiver at the Quaternary Isotope Laboratory at the University of Washington. Professor Stuiver's job was to provide an estimate of the date at which each specimen was deposited (died). This work is done in two stages. In the first, Stuiver uses a technique called *radiocarbon dating.* This gives him first approximations to the dates at which the specimens died. In the second stage, he uses a calibration technique to improve the approximation. Statistical techniques play important roles in both stages.

At the first stage, Stuiver converts Sieh's specimens to the "purest carbon dioxide in Seattle" and then measures their level of radioactivity. He wants to find out how much of the radioisotope, ^{14}C (radiocarbon), is present in each specimen. He follows a technique set down in 1945 by Professor William Libby at the University of Chicago. Libby knew that living matter, such as a tree, contains a near constant level of ^{14}C during its lifetime. Once the tree dies, however, the ^{14}C decays into another element at a known rate. This is shown in Figure 2. For example, the ^{14}C will be reduced to a half of what it was in 5,568 years and to a fifth of what it was in about 13,000 years. (In Figure 2, the 5,568 is indicated by the solid vertical line. It is referred to as the *half-life.*) So, as Libby saw, it will be possible to find the date of a specimen's death if it is known how much ^{14}C it had originally and how much it has now. The original quantity is impossible to get directly. So at this first stage, Stuiver makes the assumption that the amount of ^{14}C in living material has remained about the same across time and uses a standard material (oxalic acid) to get an estimate of how much radiocarbon there would have been in each of Sieh's specimens at the time of its death. This gives him a proportion. If, for example, the proportion is 0.9, then, using the curve of Figure 2, the time elapsed is approximately 850 years and the corresponding date at which the specimen died is $1987 - 850 = 1137$. The year 1137 would be Stuiver's first approximation to the date of the specimen's death. It is referred to as the *radiocarbon date* of the specimen. It should be mentioned that a variety of corrections, for example, for background radiation, are also applied in the course of determining a specimen's radioactivity and radiocarbon date.

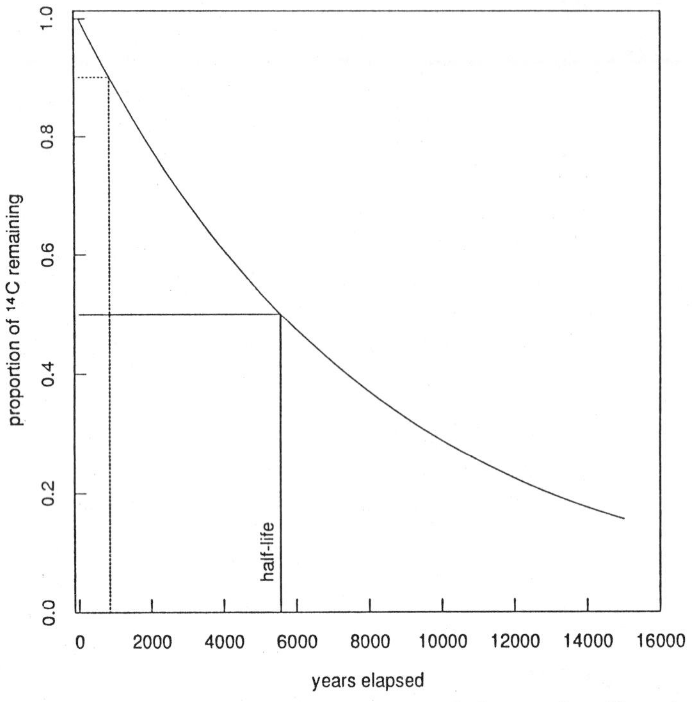

Figure 2 *Plot showing the exponential decay of radioactivity and half-life of radiocarbon. The solid vertical line indicates years passed corresponding to a proportion of 0.5. It provides the half-life of 5,568 years. The dotted vertical line indicates years passed corresponding to a proportion of 0.9.*

CALIBRATION

Professor Stuiver and others have carried out a variety of radiocarbon datings on specimens (tree rings) of known date. They have found that ^{14}C activity in the atmostphere has not remained precisely constant, as Libby initially assumed, but has fluctuated to an extent. Knowing both the radiocarbon and calendar dates of these specimens, the researchers were able to prepare a *calibration curve* relating the two. At the second stage of his work, Stuiver employs such a curve to determine an improved estimate of the calendar date of a given specimen. Figure 3 presents this calibration curve (see Stuiver and Pearson, 1986). For a given radiocarbon date one can read off a calendar date. Suppose one has a specimen with a radiocarbon date of A.D. 1100, then the corresponding calendar date is about A.D. 1200 (see the horizontal and vertical lines in the figure). The calibration operation has been crucial, changing the date by about 100 years.

In his work Stuiver has to deal with measurement errors and to compute estimates of unknown quantities. He also wishes to provide measures of the uncertainties of his estimates. Statistics has a variety of techniques for addressing these problems.

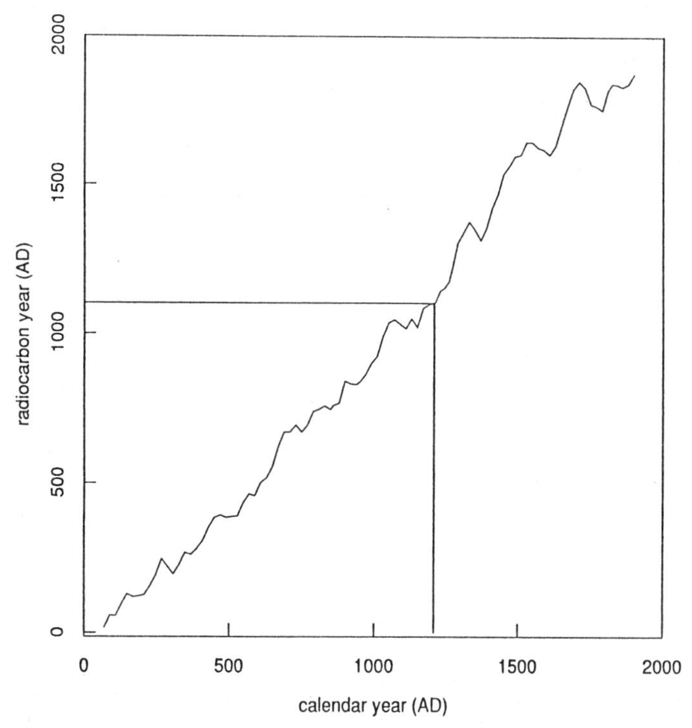

Figure 3 *Calibration curve indicating radiocarbon years and corresponding calendar years. The vertical solid line near 1200 gives the calendar year corresponding to a radiocarbon year of 1100.*

THE STATISTICAL APPROACH

Foremost among the concepts fundamental to the statistical approach to scientific problems is the notion of *distribution*. Supposing that it makes sense to talk of probabilities attached to a circumstance of interest, then the distribution of a numerically valued quantity is the function giving the probability that the quantity takes on a value not greater than a specified number. Figure 4 gives two examples of cumulative distribution functions, a normal and a Weibull. The family of normal frequency distributions have a specific shape—they are symmetrical and have one hump—and they are useful at describing frequency distributions whose observations may represent the sum of many independent contributions such as stature, or scores on achievement tests. The Weibull distributions are a class that have long right-hand tails and have been especially useful for describing the results of life tests, such as time to failure of light bulbs and fatigue tests. More generally, Weibulls can describe the distribution of the time until the next event in a series happens. Figure 4 gives as examples cumulative distributions, and the hump of the normal family is represented by the steep slope near the mean. The long tail of the Weibull is represented by

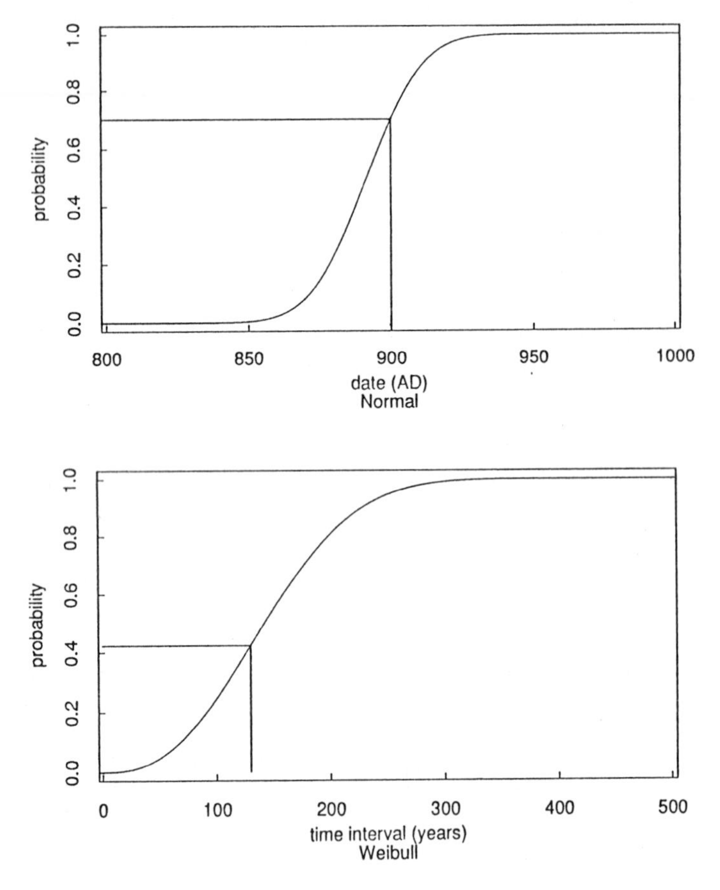

Figure 4 *Examples of distribution functions for two particular distributions (a normal and a Weibull). The curve gives the probability of not exceeding a specified value along the x-axis. In the top graph, the probability corresponding to not exceeding 900 is indicated. In the lower graph, the probability of not exceeding 130 is indicated.*

the slow rise at the right of the figure. From the top graph of Figure 4 one may read that the probability is about .70 of a value, (in this case a date) occurring that is no greater than 900. From the bottom graph, one reads .42 for the probability of a result (in this case a time interval) no greater than 130. Distributions are employed in the construction of *statistical models* (manipulable probabilistic descriptions of situations of concern). With a statistical model, one can address a host of scientific questions in a formal manner.

Distributions generally come in families, individual members of which are labeled by *parameters.* The top graph of Figure 4 illustrates a case of the *normal* family with parameters 891.7 and 15.7, while the lower illustrates a case of the *Weibull* with parameters 2.55 and 164.4. (These particular parameter values are used in calculations with the data later in this essay.) For the normal distribution the first parameter whose value is 891.7 is the mean or average value, and the second parameter is the standard deviation—a measure of spread. (See the essay by Zabell for more detail about the standard deviation.) *In some contexts the term standard error is used for the standard deviation of a measure such as a mean or some other observation.* Standard error *will be used in the remainder of this essay.* We will not describe the parameters of the Weibull distribution.

In statistical work with data a central concern is choosing an appropriate distribution to employ. ***Probability plotting*** is one technique for discerning a reasonable family. In probability plotting one graphs on special graph paper an estimate of the distribution function versus a member of a contemplated family. If the family is reasonable, the points plotted will lie near a straight line. Figure 6 (discussed later), provides an example of a Weibull probability plot for the data of concern to Professor Sieh.

Once a family of distributions has been selected, then there is a need to know the parameter values for the best-fitting particular member of the family to the data. This involves ***parameter estimation.*** Parameter estimation is often conveniently approached via the ***likelihood function.*** The likelihood function of a hypothesized probability model and given data set is a particular function of the parameters of the model that measures the weight of evidence for the various possible values of the parameters. Figure 5 (discussed below) provides an example with the unknown being the calendar date of one specimen of interest.

After an estimate of a parameter has been found, it is usual to provide some indication of its uncertainty. One particularly convenient means of doing this is via a ***95% confidence interval.*** These are numerical intervals constructed in such a fashion that over the long run 95% of them will actually contain the true parameter value. The interval of dashes within the curve in Figure 5 is an example of such an interval.

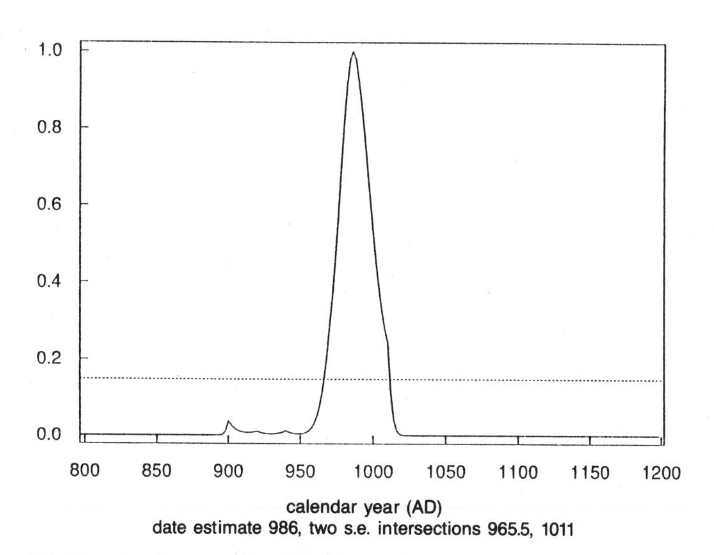

Figure 5 *A plot of the likelihood function corresponding to a specimen of radiocarbon dated 891.7 with a standard error of 15.7. The central peak corresponds to an estimate of the calendar date of the death of the specimen, here* A.D. *986. The dotted line that intersects the curve indicates a 95% confidence interval for the calendar date of the specimen.*

The above statistical concepts* play a role in the analyses described in this essay.

ESTIMATED DATES

Stuiver found one of Sieh's recent specimens to have a radiocarbon date of 891.7, with a standard error of 15.7. Figure 5 provides the likelihood function for the calendar date of deposition of this specimen. In computing this likelihood function the statistician takes into account that both the specimen's radiocarbon date estimate and the calibration curve are subject to measurement errors with approximate normal distributions. The radiocarbon date error depends, in part, on over how long a time period the specimen's level of radioactivity was measured in the laboratory. The calibration curve error depends, in part, on how many known-age items were included in its construction. The date for which the likelihood is largest here is A.D. 986 (see Figure 5). This particular specimen was selected by Sieh to provide a date between the earthquakes that he has labeled I and N in Table 1. By using an interval of twice the standard error on each side of the estimate, we get approximately 95% confidence intervals as shown in Table 1. The 95% confidence interval for the specimen's calendar date runs from A.D. 965 to A.D. 1011. This interval corresponds to the points where the dotted line in the figure intersects the curve in Figure 5. We have no confidence interval for the first entry, 1857, because it is part of recorded history.

In practice there is sometimes an added difficulty. The calibration curve is not steadily increasing as a function of the calendar date. Wiggles appear in it due to things like solar magnetic field disturbances, changes in the Earth's magnetic field, and the measurement error already referred to. The wiggles mean that sometimes one cannot associate a given radiocarbon date with a unique calendar era. To sort out the eras, one needs supplementary information.

INTER-EVENT TIMES

Sieh (1984) lists the following estimated calendar dates for 10 earthquakes at Pallett Creek: 1857, 1720, 1550, 1350, 1080, 1015, 935, 845, 735, and 590. (These are given in Table 1, as well as twice their associated standard errors.) Only one of these dates was available historically, namely, 1857. The other dates were derived by Sieh by interpolation between the estimated dates of the various specimens he selected in the course of his excavations.

At this stage of his study, Sieh turned to a statistician for assistance in inferring the probabilities of future earthquakes. (The radiocarbon daters had turned to statisticians earlier in the development of their estimation procedures.) In

*See Nelson (1982) for an explanation of the ideas of probability plotting, the normal and Weibull distributions, and related statistical concepts.

Table 1 Estimated dates and twice their standard errors for historical earthquakes at Pallett Creek. [The event labels and values are those of Sieh (1984).]

Event	*Date,* A.D.
Z	1857
X	1720 ± 50
V	1550 ± 70
T	1350 ± 50
R	1080 ± 65
N	1015 ± 100
I	935 ± 85
F	845 ± 75
D	735 ± 60
C	590 ± 55

Source: Sieh (1984).

the statistical approach to the problem of probability estimation, one seeks a distribution function for the series of times between the events. From the smallest to largest these times are: 65, 80, 90, 110, 137, 145, 170, 200, and 270 years, with 131 years now passed in 1988 since the 1857 event. The statistician sets out to determine a statistical model for these values.

The Weibull family has often been found applicable for the lifetimes of items subject to destruction and for other related phenomena. A Weibull probability plot was prepared for Sieh's data. It is given in Figure 6. The vertical bars correspond to the dating errors of the corresponding interevent times. If the Weibull is adequate for describing the distribution of times between earthquakes, then the points plotted should fall near a straight line. For reference, a straight line has been included in the figure. The Weibull assumption appears reasonable here.

RISK ESTIMATES

Many people are interested in such questions as: What is the probability of a large earthquake in the Los Angeles area in the next 5 years? In the next 10 years, and so on? These probabilities (risks) may be estimated once one has a distributional form for the times between earthquakes. Figure 7 provides preliminary estimates of risk probabilities, employing the Weibull distribution referred to and using the data of Table 1. From the figure one sees, for example, that the probability of a large earthquake in the next 30 years given that the last earthquake was 131 years ago may be estimated by .32 (that is, there is a 32% chance of one occurring). The dashed lines in the figure provide an indication of the uncertainty in the fitted probability values. They correspond to a 95% confidence interval. By following the horizontal lines of the figure one is led to a lower limit of .18 and an upper limit of .50 for the probability

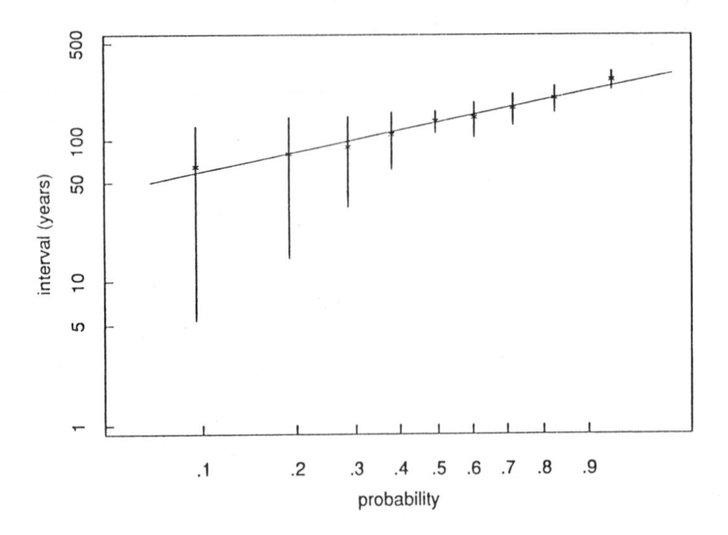

Figure 6 *A probability plot to assess the reasonableness of the Weibull distribution for the intervals between earthquakes at Pallett Creek. The points plotted correspond to the observed intervals. The vertical bars indicate plus and minus twice their standard errors. If the distribution is reasonable the points should fall near a straight line. For reference a fitted line has been added.*

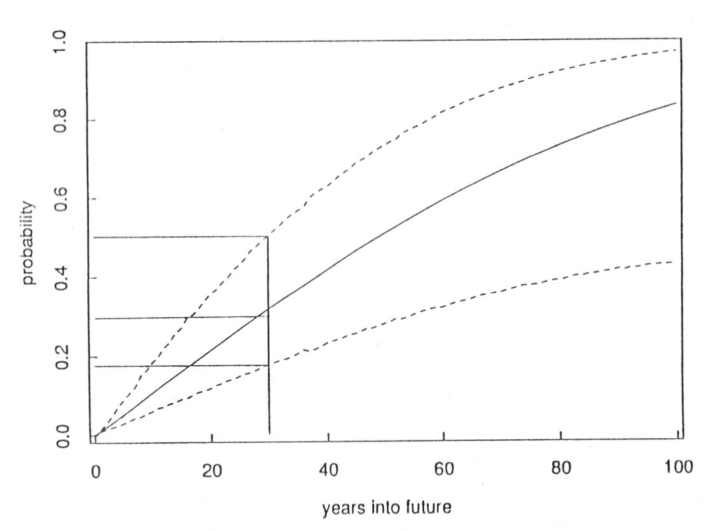

Figure 7 *An estimate of the probability of a future earthquake occurring at Pallet Creek within the indicated number of years. The dashed lines give the upper and lower values of a corresponding 95% confidence interval for the probability. The horizontal lines provide these values for the probability of an earthquake in the next 30 years.*

of an earthquake occurring in the next 30 years. This result may be used by insurers, engineers, and planners in their work.

INSURANCE PREMIUMS

Suppose one wishes to set aside funds to cover the cost of rebuilding a facility that might become damaged in an earthquake in the coming year. The fair premium to cover the rebuilding, were an earthquake to take place, per thousand dollars of cost, is given by a thousand times the probability of an earthquake occurring in the coming year. Using the fitted Weibull, the estimated probability of an earthquake in the coming year is .0108 and the premium works out to be $10.80. (Of course, insurance companies actually "load" their premiums by adding amounts to cover costs, to allow profits, and to protect themselves against extreme catastrophes, so they would charge more than $10.80.)

CONTRIBUTIONS OF STATISTICS TO THIS PROBLEM

The desired end product of a seismic risk study is a probability. So statistics is bound to enter, as statistical distributions are basic to the estimation of probabilities. In the study just described, the tool of radiocarbon dating was crucial. Researchers in that field have long recognized the importance of good statistical techniques. As H. A. Polach (1976) has said, "The application of sound statistical methods has become a radiocarbon dater's 'bread and butter.' "

It is also worth quoting Harold Jeffreys (1967), one of the most important seismologists and statisticians of this century. He has said that "An estimate without a standard error is practically meaningless." This refers to the statement of conclusions. Providing standard errors is one tool for this, confidence intervals are another. These are both central concepts of statistics. Remember that the 95% confidence interval for the probability of a large earthquake at Pallett Creek in the next 30 years runs from .18 to .50.

CONCLUSION

Science proceeds by building on itself. In the work described, specimens of known age (tree rings) are employed to construct a calibration curve that is employed in dating specimens of unknown age. Science uses statistical concepts to address problems of estimating unknowns, to validate assumptions, and to quantify uncertainties in the inferences made.

PROBLEMS

1. Use the curve of Figure 2 to read off the years elapsed for the radiation to drop to a quarter of its initial value.

2. Use the curve of Figure 3 to read off the calendar year corresponding to a radiocarbon year of 500.

3. Use the curve of Figure 3 to find a radiocarbon year that corresponds to several calendar years rather than to a unique year. Comment on this phenomenon.

4. There is a bump around the year A.D. 900 in the curve of Figure 5. What do you think its source is? (Hint: Consider Figure 3.)

5. What is the approximate fair insurance premium to pay to cover $50,000 worth of damage due to an earthquake that might take place in the next 20 years? (Hint: Read a probability estimate from Figure 7.)

6. Evaluate the times between successive events for the earthquakes listed in Table 1. Does there seem to be any structure in the sequence of values?

REFERENCES

D. R. Brillinger. 1982. "Seismic Risk Assessment: Some Statistical Aspects." *Earthquake Prediction Research* 1:183–195.

H. Jeffreys. 1967. "Seismology, Statistical Methods in." In *International Dictionary of Geophysics,* K. Runcorn, ed. London: Pergamon Press, pp. 1,398–1,401.

W. Nelson. 1982. *Applied Life Data Analysis.* New York: Wiley.

H. A. Polach. 1976. "Radiocarbon Dating as a Research Tool in Archaeology: Hopes and Limitations." In *Proceedings of a Symposium on Scientific Methods of Research in the Study of Ancient Chinese Bronzes and Southeast Asian Metal and Other Archaeological Artifacts,* N. Barnard, ed. Melbourne, Australia: National Gallery of Victoria, pp. 255–298.

K. E. Sieh. 1984. "Lateral Offsets and Revised Dates of Large Prehistoric Earthquakes at Pallett Creek, Southern California." *Journal of Geophysical Research* 89:7,641–7,670.

M. Stuiver and G. W. Pearson. 1986. "High-Precision Calibration of the Radiocarbon Time Scale, A.D. 1950–500 B.C." *Radiocarbon* 28:805–838.

Very Short Range Weather Forecasting Using Automated Observations

Robert G. Miller *National Weather Service*

Pilot Joe Lang was using his medical-evacuation helicopter to take an accident victim from the beltway to a nearby trauma center, and after 10 miles of his 20-mile trip, the weather changed at his destination, leaving him no visibility to make a landing. Precious time was lost finding an alternate landing site at the last moment because there had been no short range forecast. Helicopter pilots want and need help in preventing this loss of time and the life-threatening disaster of a crash.

This essay describes a statistical effort to produce very short range weather forecasts to help helicopter pilots and others. The work was conducted in conjunction with the automated weather observing system (AWOS) currently under test. An AWOS observation is made at an unstaffed airfield by taking minute-by-minute readings from meteorological instruments (sensors that measure temperature, wind, and so forth). The most recent 30 minutes of these readings are sent electronically to a computer programmed to emulate human observations in creating a weather observation. There are three alternative ways to make short-range forecasts of, say, 10–120 minutes:

1. *Persistence forecasts.* These are forecasts that the current conditions will prevail into the short-term future. This method provides good results and is not easy to beat. We know that weather changes infrequently over short time periods, so persistence forecasting is bound to be rather accurate.
2. *Human judgment.* Human forecasts are an important alternative because all current and past weather information is at the disposal of the forecaster. This information includes observations from weather radar and satellites, spatial and temporal weather conditions at the surface and aloft, large-scale prognoses, and, of course, the local AWOS observation. Combined with the forecaster's experience and training, this information looks like the best source for 10–120-minute forecasts. Experience has shown however, that such forecasts rarely show any gain over persistence forecasting for less than 120 minutes.
3. *Statistical prediction.* The statistical approach employed here, that of basing forecasts on empirical data, represents another alternative. It has the advantages of using only the most recent locally available AWOS observation as input, of being easily automated, and of possessing a rapid response time. The approach is founded on almost a million past observations that generate estimated relationships.

Because short-term forecasts based on human judgment that show improvement over persistence are hard to provide, can statistical prediction help?

CREATING PREDICTION EQUATIONS

For purposes of illustration, we describe a statistical approach for making a 10-minute prediction of visibility that is directly compared with persistence. The approach is called *regression analysis* (or two-group discriminant analysis; see Tatsuoka, 1971). (See also the essay by Howells on the discriminant function.) It is a popular way of estimating the value of an unknown variable that is given in terms of weighted values of known variables. (In particular, the weights are determined in such a way as to minimize the sum of squares of the differences between the estimates and the variable's true values.) We choose these weights by examining situations where we do know the values of the usually unknown variable.

Visibility is measured in miles and fractions of miles by a visibility sensor. The AWOS visibility is derived from the 10 most recent visibility sensor readings. We could attempt to predict this derived value but instead it turns out to be more convenient to report in which of the following six categories of 10-minute visibility the observation and the prediction lie:

- 0 up to but not including 1/2 mile
- 1/2 up to but not including 1 mile

- 1 up to but not including 3 miles
- 3 up to but not including 5 miles
- 5 up to but not including 7 miles
- 7 miles or more

In a given forecast situation, an AWOS observation is represented by a series of 1s and 0s. For example, the weather element wind speed has five velocity categories, only one of which can occur in the observation. That category gets a 1 and the other four get 0s. Similarly, wind direction has eight direction categories of which only one can occur in the observation. Altogether there are 26 weather elements represented by 166 categories. A given observation of the 166 categories has exactly 26 1s and 140 0s. This form of representation has two important purposes. First, although some weather elements are ordered, they are not on a numerical scale (visibility is an example in which there are no numeric values for the category "unlimited" and "7 miles or more"). Second, the mathematical processing of the categorical data is done very efficiently by logical computer operations, replacing slower arithmetical operations. (For example, it takes a microcomputer 20 times longer to obtain the multiplication of two real numbers than for a logical operation between two integers where each is representing many 0/1 numbers. All told, the comparison works out to be about two orders of magnitude in computation time and about one order of magnitude in storage space in favor of the 0/1 scheme.)

The statistical analysis is a regression procedure that selects 30 predictors from the 166 variables (see Draper and Smith, 1981, and Miller, 1962). A few examples of potential variables whose categories are used as predictors are: lowest height observed by cloud sensor in one minute, pressure, wind direction, and precipitation amount. On the basis of success in estimating the outcome of 818,953 previous observations, the statistical procedure selects the 30 predictors, or categories, and also computes weights for them that would use their outcome to good effect. The overall method produces a separate estimate or score for each of the categories of 10-minute visibility listed above. Ideally the outcome would be a 1 for the category that is going to happen and a 0 for all others, but actually the numbers vary from somewhat below 0 to somewhat above 1. Each category has its own equation.

The first regression estimate gives us for 10-minute visibility the number associated with its first category listed above—0 up to but not including 1/2 mile:

Regression estimate =

0.047 + (−0.001) × (visibility sensor 7 miles or more)
+ 0.003 × (visibility sensor 1 mile to less than 3 miles)
+
.
.
.
+ (−0.009) × (precipitation accumulation 0.002 − 0.100 in 1 minute)

The terms to the right of the equal sign are algebraically added according to whether the corresponding predictor condition within the parentheses is occurring or not. For example, we start with 0.047, then add –0.001 if the visibility sensor is 7 miles or more; but if the visibility is not 7 miles or more, the term –0.001 is deleted. The remaining 28 coefficients are added or not added in a similar fashion depending on the observed condition of the predictor. The quantity obtained as the regression estimate is an index from which the final visibility forecast is made and will be described later.

Recall that regression analysis computes equation weights that minimize the sum of squares of the error between the forecast value and the condition being forecast. Table 1 sums up how well the two methods perform. It compares the sums of squares of deviations between the observed and forecast values for each category and for each method with the sums of squares using just the mean performance for each category. Thus, persistence forecasting had a sum of squares of deviations for the 0 to 1/2 mile category that was 50.2% as big as that for the sum of squares of deviations from the average. The complement, 100 – 50.2 = 49.8, is called the *predictability*.

The persistence percentage of total predictability was obtained as follows: Persistence probability forecasts were obtained by applying a regression approach similar to the statistical method described above, except that only persistence predictors (the six categories of visibility at time 0 listed above) were selected for inclusion into their (persistence's) equations. This is shown in column 1. The improvement of regression over persistence is given in the last column of Table 1.

The amount of improvement shown by regression over persistence indicates that there is a sizable contribution being made by the predictors in the regression equation both in absolute percentage of total predictability and over and above what the persistence terms in the equation are contributing. These quantities indicate that we can expect better forecasts when the regression equations' index values are applied to the final step in the forecast process. We turn now to the actual process used and how well it works.

COMPARATIVE RESULTS

The goal of this entire effort is to utilize the index values produced by the regression analysis and to predict the category within which visibility will be observed to occur 10 minutes hence. Numerous approaches could be taken. However, the category with the highest index value has been found to produce the largest number of correct forecasts. Unfortunately, this criterion tends to provide little chance of forecasting categories that occur less often and favors those that occur more often. This phenomenon is most evident as the projection time of the forecast is extended. What is desired by both practicing meteorologists and operational users is a balance in the frequency of *each* category—balance in that the forecast frequencies agree closely with the observed frequency for each category of the event. For example, visibilities of 0 to 1/2 mile occur only

Table 1 Percent of total predictability

Predictor Category (Miles)	*Persistence (%)*	*Regression (%)*	*Regression improvement over persistence (%)*
0 < 1/2	49.8	54.8	5.0
1/2 < 1	31.5	40.6	9.1
1 < 3	56.5	64.3	7.8
3 < 5	54.9	62.4	7.5
5 < 7	44.5	53.3	8.8
7^+	83.9	87.6	3.7

about 0.4% of the time while visibilities of 7 miles or more occur about 75.5% of the time. Our scheme must predict with roughly these percentages or we will fail to meet the standards of balance that have evolved within the meteorological profession.

A method has been devised by Klein et al. (1959) to satisfy the variability condition preferred by meteorologists and users. We shall not describe it here, but will turn instead to the proof of the pudding.

We can evaluate the stability of our statistical procedure by applying the forecasts to a test sample of data. Tallies for a sample of 369,802 10-minute visibility forecasts are given in Table 2 for the statistical procedure. Table 3 shows the corresponding set of results obtained using persistence.

The results from the test-sample obtained in this effort are extremely gratifying. The number of correct forecasts for the statistical method is obtained from Table 2 by starting with 1,032 and going down the diagonal 1,032 + 1,549 + 14,562 + · · · + 274,368 = 345,240. From Table 3, persistence had 1,061 + 1,382 + 13,926 + · · · + 273,849 = 342,736 correct forecasts. Thus, the statistical method gave 2,504 more correct forecasts than did persistence. Since persistence missed in 369,802 − 342,736 = 27,066 forecasts, regression has succeeded in correcting more than 9% of those misses. Another encouraging fact is that the statistical scheme changed 180 (1,450 − 1,270) of those situations where the visibility was 0 to 1/2 mile at forecast time and only had 29 fewer hits in that category than persistence did. The percentage of correct forecasts of this category is 1,032/1,270 or 81.3%, which is better than the 1,061/1,450 or 73.2% for persistence.

Obviously forecasting high visibility and observing low visibility is undesirable. Table 3 has 13,297 persistence situations below the diagonal (the undesirable forecasts), while regression (Table 2) has 11,389, or 1,908 fewer. The crucial area of three or more below the diagonal in this area of the table is even more impressive, with 153 for persistence and 85 for regression, or 68 fewer very bad forecasts for regression.

The final conclusion is that statistics has succeeded in improving on persistence forecasts of a very difficult meteorological element, visibility, and has done so at the very short range of 10 minutes. The value of correctly forecasting changing conditions, where persistence by its very nature does not make such forecasts, is an accomplishment that cannot be overemphasized.

Table 2 Statistically based forecasts versus observations for 10-minute visibility predictions on a test sample of 369,802 observations

Forecast Category (Miles)	*Verifying Observations (Miles)*						
	0 < 1/2	*1/2 < 1*	*1 < 3*	*3 < 5*	*5 < 7*	*7⁺*	*Totals*
0 < 1/2	1,032	162	52	13	6	5	1,270
1/2 < 1	261	1,549	657	90	31	16	2,604
1 < 3	52	476	14,562	2,884	196	103	18,273
3 < 5	5	34	2,341	27,660	4,312	382	34,734
5 < 7	1	3	133	3,405	26,069	4,264	33,875
7^+	0	0	76	351	4,251	274,368	279,046
Totals	1,351	2,224	17,821	34,403	34,865	279,138	369,802

Table 3 Persistence versus observations for 10-minute visibility predictions on a test sample of 369,802 observations

Forecast Category (Miles)	*Verifying Observations (Miles)*						
	0 < 1/2	*1/2 < 1*	*1 < 3*	*3 < 5*	*5 < 7*	*7⁺*	*Totals*
0 < 1/2	1,061	249	110	17	7	6	1,450
1/2 < 1	197	1,382	590	40	11	5	2,225
1 < 3	80	530	13,926	2,998	207	112	17,853
3 < 5	8	47	2,828	26,673	4,251	564	34,371
5 < 7	2	11	243	4,205	25,845	4,602	34,908
7^+	3	5	124	470	4,544	273,849	278,995
Totals	1,351	2,224	17,821	34,403	34,865	279,138	369,802

PROBLEMS

1. Joe Lang can't land his helicopter if the visibility is, say, less than 1 mile. If he had followed the statistical forecasting approach during the sampling period covered in Tables 2 and 3 how many times would he have had to cancel his flights? How many times would he have had to cancel if he had used persistence? He has to divert to another airport when the forecast is incorrect. Compare the two approaches in light of the actions required, and state a case for choosing one method over the other.
2. Why do you think users of weather forecasts represented by categories prefer them to be issued with the same frequency as they are observed? Is this requirement reasonable? Give the pros and cons as you see them.
3. Does it make you uncomfortable to realize that very short range forecasts might best be issued by a computer program and not from a human being's judgment? Express your opinion.
4. Why do you suppose forecasts are not accurate enough to satisfy our needs? Is it that our demands are too high, our understanding of the atmosphere too inadequate, our data too incomplete, or our analytical methods too limited?

REFERENCES

N. R. Draper and H. Smith. 1981. *Applied Regression Analysis.* New York: Wiley.

W. H. Klein, B. M. Lewis, and I. Enger. 1959. "Objective Prediction of Five-Day Mean Temperatures during Winter." *Journal of Meteorology* 16: 672–681.

R. G. Miller. 1962. *Statistical Prediction by Discriminant Analysis.* Meteorology Monograph No. 25. Boston: American Meteorological Society.

M. M. Tatsuoka. 1971. *Multivariate Analysis: Techniques for Educational and Psychological Research.* New York: Wiley.

Statistics, the Sun, and the Stars

Charles A. Whitney *Harvard University*

The visible stars appear to be scattered at random in space, so it is natural that astronomers should have turned to statistical methods and probabilistic arguments. But this is a relatively recent development. Babylonian and Greek astronomy was based on cycles, and the concepts of probability were rarely used in modern astronomy before the twentieth century.

An early use of the idea of randomness and probability in astronomy was John Michell's mathematical demonstration, in the eighteenth century, that most stars that appear to be very close together in the sky actually are close together in space. William Herschel, a contemporary of Michell, discovered an abundance of very close pairs of stars—*double stars* they were called—but he had no way of telling whether the stars really were paired or merely appeared to be. Herschel hoped that the individual stars in a pair were quite far apart and that they accidentally lay along the same line of sight; from observing such pairs he hoped to detect the motion of the earth about the sun and, ultimately, to determine the distances of the stars from earth. But Michell computed the likelihood that such apparent pairing could arise in the numbers found by Herschel if the stars were scattered at random through space; the probability was so minute that Michell believed the stars to be physically coupled. He was later proven correct by Herschel himself. Herschel found many pairs in which the members were rotating about a common center of gravity, and he was thus led to the first demonstration of the application of Newton's concept of universal gravitation outside the solar system.

Herschel also attempted to map our Milky Way galaxy by studying how the number of stars seen in a particular direction increased as he went to fainter and fainter limits. For lack of better information, he assumed that all stars would appear the same at a standard distance, and he found that the sun appeared to be at the center of the Milky Way. Astronomers now recognize that this is an illusion—the space between the stars is littered with dust and gas that dim the light of distant stars and at the same time alter its color. Modern data and statistical methods have led to a model in which the Sun is displaced from the center and is embedded in one of our galaxy's spiral arms.

On a grander scale, recent studies have shown that the galaxies are distributed throughout space in a highly irregular pattern interspersed with vast empty regions resembling bubbles. Understanding this pattern is a major task of current cosmology.

Generally speaking, the statistical arguments now used by astronomers in their attempts to unravel the causal connections woven into the sky are of two classes: first, statistical analyses of data, and, second, physical theories based on statistical or probabilistic concepts. An example of each will be presented.

ANALYSES OF DISTURBANCES IN THE SOLAR ATMOSPHERE

The solar atmosphere seethes with activity, and some disturbances, the gentler ones, are especially amenable to statistical analysis. The solar weather, so to speak, is not altogether chaotic, and astronomers are anxious to glean what they can about the regularities of the pattern because such regularities invariably assist the construction of theoretical explanations.

The vapors of the solar atmosphere are intensely hot, so hot, in fact, that they radiate visible light. Yet, their heat is far from uniform; a telescope reveals a welter of evanescent detail that surges and disappears from place to place within brief minutes. Disturbances are strewn in an irregular pattern—the hot and cool areas cover hundreds of miles and their outlines are roughly hexagonal. This pattern is called *granulation* and the hexagon-shaped elements, called *granules,* are evidently bubbles of hot gas welling up from the interior, carrying heat from the center and disturbing the delicate outer layers. Even the most powerful telescope cannot penetrate beneath the solar atmosphere, so astronomers rely on mathematical analysis to assess the solar interior. But this analysis requires an observational check. The detailed nature of the atmospheric fluctuations provides such a check, and it also permits astronomers to probe the interior of the Sun, similar to the way a seismologist may probe the interior of the earth.

BRIGHTNESS

Two types of measurements can be made to infer the structure: the brightness and the velocity of the gas. The earliest studies showed that the pattern of brightness changed drastically every five minutes or so. Quantitative data have been obtained since from series of photographs exposed briefly every 10 or 20 seconds. The exposures are typically one-hundredth of a second, and the best series are taken immediately after sunrise, when the ground is cool and the air is steady. These series often cover several hours, and astronomers at the polar regions during the seasons of the midnight sun have been able to obtain some sequences lasting many dozens of hours.

Comparisons of individual photographs (Figure 1 is an example) separated by longer and longer intervals reveal changes in brightness associated with time lapses. The cross-correlation of the patterns on different films is determined in the following way: a line is specified on the sun's surface, and the intensity of the light at points along the line is determined on each photograph. (Those who are not interested in the measure of similarity or who are familiar with the correlation coefficient may want to skip past the following material to the heading "A Possible Model.")

Table 1 represents five points along a line on the sun's surface. The first row of numbers gives initial brightnesses. These are measured in a standardized manner so that brightness at a point is measured as a departure from the mean in units of the original variability. This standardizing gives these numbers the properties that they average to 0 and their squares average to 1. (In this table, these averages do not work out exactly because of rounding off.) The second row gives standardized brightnesses a little later in time, but at the same positions. The third row gives standardized brightnesses a few minutes later.

Note that the second row is almost the same as the first, but the third row differs a great deal from the first. One way to measure similarity between two sets of five numbers is to multiply the two standardized numbers of each pair, add the five products, and divide by five to get the average. The underlying idea is that this average will be near 1 if there is high similarity, near 0 if there

Table 1 The idea of correlation

			Points		
Time	*1*	*2*	*3*	*4*	*5*
1	1.5	0.5	0	− 0.5	− 1.5
2	1.4	0.7	0	− 0.7	− 1.4
3	1.0	− 1.4	0.4	− 1.0	1.0

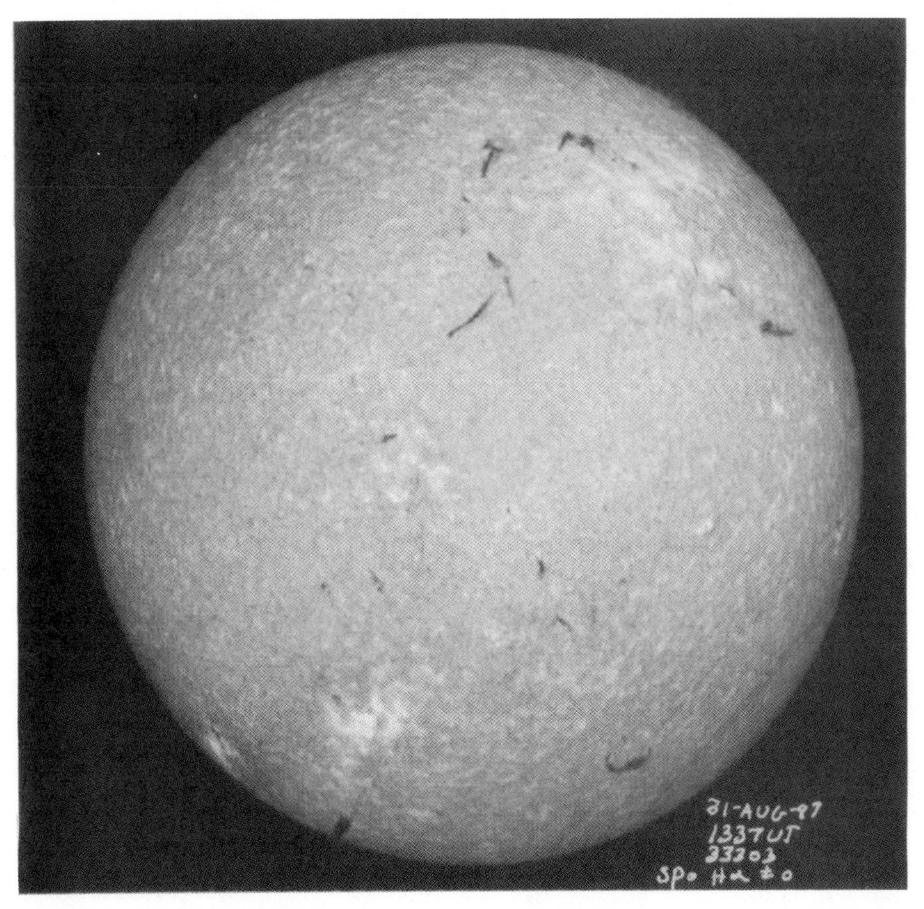

Figure 1 *Photograph of the sun's surface showing fine details that have been subjected to statistical analyses of various sorts. The light and dark areas reveal regions of different magnetic intensity. Source: National Solar Observatory.*

is little connection, and near -1 if there is high dissimilarity. The average product is called the *correlation coefficient.*

Let's first look at an extreme case of similarity; when we compare the first set of five numbers with itself, we should get perfect similarity. The correlation coefficient is

$$\tfrac{1}{5}[1.5 \times 1.5 + 0.5 \times 0.5 + 0 \times 0 + (-0.5) \times (-0.5) + (-1.5) \times (-1.5)] = 1,$$

as we said it would be. The similarity of a set of such standardized numbers with itself when measured this way always gives a value of 1, just as in this example. The correlation coefficient can range from $+1$ to -1. If we can't predict the values at one time from those at another any better than we would by guessing, the correlation coefficient gives the value 0.

Correlating the first set of brightnesses with the second, we get

$$\tfrac{1}{5}[1.5(1.4) + 0.5(0.7) + 0(0) + (-0.5)(-0.7) + (-1.5)(-1.4)] = .98.$$

This pair is very highly correlated, although, of course, slightly less than the original numbers with themselves.

Correlating the first set with the third gives us

$$\tfrac{1}{5}[1.5(1.0) + 0.5(-1.4) + 0(0.4) + (-0.5)(-1.0) + (-1.5)(1.0)] = -.04,$$

a slightly negative value, but not far from 0. The example is primarily for illustrative purposes, as the correlation ordinarily stays positive.

If every standardized brightness were replaced by its negative, we would find a coefficient of −1 between the original and the new values. Let's try it with the first set of numbers:

$$\tfrac{1}{5}[1.5(-1.5) + 0.5(-0.5) + 0(0) + (-0.5)(+0.5) + (-1.5)(1.5)] = -1.$$

Although the correlation coefficient ranges between −1 and 1, other ways of standardizing are possible. But the −1, 1 interval is conventional and convenient.

A Possible Model. This correlation will be positive if bright points on one film correspond with bright points on the other film—as will be the case if the films were separated by a very short interval of time. The correlation will be negative if brighter points on one correspond to darker points on the other, and if we cannot forecast one set from another better than guessing will do, the correlation will be 0.

As the time interval increases, the correlation has been found to decrease steadily toward 0 without actually going negative. The value of the correlation is reduced to about 1/2 in an interval of five minutes and to somewhat less than 1/4 in ten minutes.

A model for this process has been suggested. It assumes that at random times, averaging about five minutes apart, new granules appear and gradually cool. Each is replaced by another at the next random time. If the cooling is slow and the replacement is slow in coming, the correlation is slow in going to 0. If a replacement comes rapidly, the correlation goes to 0 rapidly. Theoretical work not given here shows that this model will create correlations that reduce from 1 to 1/2 in about five minutes on the average. This agreement with the facts lends some support to the model.

VELOCITY

The other type of measurement, velocity, has been more exciting in its consequences; it is also more difficult to obtain but a vast amount of data has been accumulated, and these data show a behavior markedly different from that of brightness changes associated with granulation. Correlations can be measured for the velocity in much the same way they are for the brightness; they also decline with time, but they dip down and become negative before they rise again to 0. In fact, a detailed plot of the velocity at a particular point on the sun's surface shows an oscillation that is quite striking in its apparent regularity. Figure 2 shows one example; it is not perfectly regular, but there is no doubt of an actual oscillation. Why are oscillations seen in the velocity, while the brightness pattern is irregular in both space and time?

The answer to this question remains incomplete, but studies of the detailed frequency composition of these oscillations are beginning to clarify the phenomenon—and to show its real complexity. These studies have shown that the oscillations are amazingly similar to waves of musical notes emitted by a violin string. Overtones are present, but they are relatively weak. In fact, over half the energy is contained in oscillations whose periods are confined between two and six minutes. As a terrestrial example, we might say that a slow motion movie of surf breaking on a beach would show about the same degree of periodicity.

Astronomers were astonished by the discovery of this regularity because they had come to think of the sun's atmosphere as the seat of mere chaos; also, the brightness variations had shown no such periodic oscillations, and astronomers had assumed that the pattern of the upward and downward motions would closely mimic the irregularity of the brightness changes.

The periodicity of the motions, and its sharp contrast with the randomness of the brightness, showed at once that astronomers were observing two different parts of the solar atmosphere; it has since been proven that the brightness variations are produced low in the visible atmosphere while the observed motions take place in the upper layers of the atmosphere. And, what is more, the nature of the motions alters with increasing height in the atmosphere—the

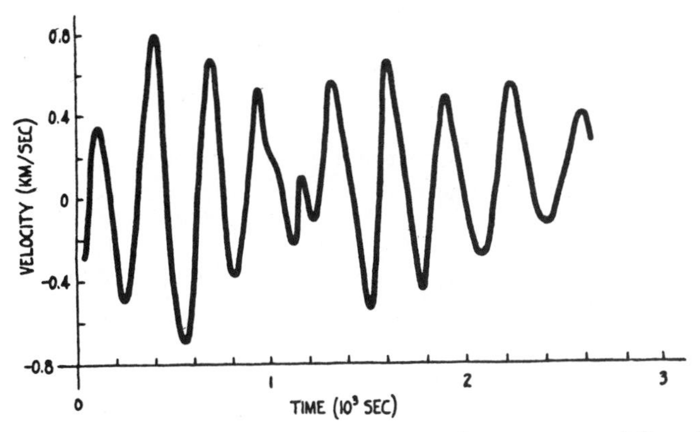

Figure 2 *Measurements of the vertical component of the velocity of gas in the sun's atmosphere are plotted here for a single point on the sun. The marked periodicity of the motions is typical of the solar photosphere, and it has been the object of many correlation studies.*

average period shortens by a factor of two, and at the greatest observable height (several thousand miles above the "surface") the velocity fluctuations become quite chaotic, resembling noisy static without any pronounced periodicity.

Why? Astronomers assume that we are witnessing the upward flight of very long "sound waves" in the solar atmosphere—waves that are generated deep in the solar atmosphere, perhaps by the rising granules. Some waves are trapped in that atmosphere, predominantly those with periods of about five minutes; waves of shorter period escape quickly to the upper levels, where they predominate. Waves of very long period die out quickly, and, in fact, they are not easily excited by the granules, so they are very weak at all levels.

Even this brief explanation makes it clear that a study of the frequencies of these oscillations may reveal several features of the solar envelope: the nature of the deeper disturbances that generate these sound waves, the rate at which waves of different periods are dissipated as they propagate, and the extent to which the solar atmosphere is capable of trapping waves of different periods.

PROBLEMS

1. What role have calculations of improbability played in the theory of double stars?
2. What are the two types of statistical arguments used by astronomers?
3. Refer to Table 1. We are interested only in the measurements taken at time 2. How does the zero entry at point 3 compare to the average brightness at all points?
4. Using Table 1, calculate the correlation coefficient for the second set of measurements with the third.
5. From the model described in the text, would you expect the correlation coefficient for times 1 and 3 to be larger than that for times 2 and 3? Did your calculation above meet your expectation? Explain.
6. Refer to Figure 2. How many oscillations fail to dip below 0? Below -0.4?
7. Draw an analogue of Figure 2 for the correlation coefficient of measurements of brightness of the sun's atmosphere made at several points on the sun.
8. The velocities presented in Figure 2 show marked periodicity. Would you expect similar graphs from other measuring points in the solar atmosphere? Explain your answer.
9. a. Describe the differences in the observed brightness and velocity patterns of gas in the sun's atmosphere.
   b. Give a partial explanation of this difference.

ACKNOWLEDGMENTS

We wish to thank the following for permission to use previously published and copyrighted material:

American Association for the Advancement of Science for Table 1 on p. 85, from Peter J. Bickel, Eugene A. Hammel, and J. William O'Connell (1975), "Sex Bias in Graduate Admissions: Data from Berkeley," *Science,* 187, pp. 398–404. Copyright 1975 by the AAAS.

American Journal of Public Health for permission to publish Table 1 on p. 12, adapted from Tables 2 and 3 from T. Francis et al. (1955), "An Evaluation of the 1954 Poliomyelitis Vaccine Trials-Summary Report," *American Journal of Public Health,* 45:5 (Part 2), pp. 1–63.

American Psychological Association for permission to publish the figures on pp. 98, 100, and Figure 9 on p. 101, from D. T. Campbell (1969), "Reforms as Experiments," *American Psychologist,* 24:4 (April), pp. 409–429. Copyright © the American Psychological Association.

American Psychological Association and Seymour Rosenberg for permission to publish the figures on pp. 135 and 139, from S. Rosenberg, C. Nelson, and P. S. Vivekananthan (1968), "A Multidimensional Approach to the Structure of Personality Impressions," *Journal of Personality and Social Psychology,* 9, pp. 283–294. Copyright © the American Psychological Association.

American Society for Quality Control for material in Lonnie C. Vance (1983), "Statistical Determination of Numerical Color Tolerances," ASQC 37th Annual Quality Congress Transactions.

Atlantic Monthly and David D. Rutstein for permission to publish the figure on p. 7, from D. D. Rutstein (1957), "How Good is Polio Vaccine?" *Atlantic Monthly,* 199, p. 48. Copyright © 1957 Atlantic Monthly Co., Boston, Mass.

Journal of Public Health for Table 1 on p. 48 from E. M. Lewitt and D. Coate, "The Potential for Using Excise Taxes to Reduce Smoking," *Journal of Health Economics,* 1, pp. 121–145.

Law and Society Association for permission to use Figure 1, p. 94, Figure 2, p. 95, and Figure 10, p. 101, from D. T. Campbell and H. L. Ross (1968), "The Connecticut Crackdown on Speeding: Time-Series Data in Quasi-Experimental Analysis," *Law and Society Review,* 3, pp. 33–53.

National Solar Observatory for photo on p. 271, Figure 1.

Operations Research Society of America for permission to publish the table on p. 191, from G. R. Lindsey (1963), "An Investigation of Strategies in Baseball," *Operations Research,* 11, pp. 477–501.

Simmons-Boardman Publishing Co. for permission to reprint the table on p. 157 from "Can Scientific Sampling Techniques Be Used in Railroad Accounting?" *Railway Age,* June 9, 1952, p. 61–64.

Springer-Verlag for permission to use tables on pp. 118, 120, 123, and 124 from F. Mosteller and D. L. Wallace (1984), *Applied Bayesian and Classical Inference: The Case of the Federalist Papers.*

University of Chicago Press for Table 1 on p. 48 from E. M. Lewitt, D. Coate, and M. Grossman, "The Effects of Government Regulation on Teenage Smoking," *Journal of Law and Economics,* 24, pp. 545–569.

Index